Melanin Chemistry Explored by Quantum Mechanics

Ryo Kishida · Susan Meñez Aspera · Hideaki Kasai

Melanin Chemistry Explored by Quantum Mechanics

Investigations for Mechanism Identification and Reaction Design

Springer

Ryo Kishida
Faculty of Dental Science
Kyushu University
Fukuoka, Japan

Susan Meñez Aspera
National Institute of Technology
Akashi College
Akashi, Hyogo, Japan

Hideaki Kasai
National Institute of Technology
Akashi College
Akashi, Hyogo, Japan

Institute for Radiation Sciences
Osaka University
Osaka, Japan

ISBN 978-981-16-1314-2 ISBN 978-981-16-1315-9 (eBook)
https://doi.org/10.1007/978-981-16-1315-9

This Springer imprint is published by the registered company Springer Nature Singapore Pte Ltd.
The registered company address is: 152 Beach Road, #21-01/04 Gateway East, Singapore 189721, Singapore

Preface

Melanin is important as a bearer of skin and hair color, that is, visual phenotype. Even within the same species, the expressions of body color observed between individuals are diverse and in some cases cause discrimination because of poor understanding. It is important to gain a better understanding of melanogenesis from atomic nuclei and electrons world.

Further, in basic and clinical medicine, understanding the pathology of dyschromia represented by oculocutaneous leukoderma and vitiligo vulgaris and melanoma, which is a malignant tumor of melanocytes, and establishing a treatment method are required.

Therefore, in dyschromia, abnormalities are mainly observed in the functions related to melanin production and transport, and in the number and shape of melanocytes. It is a designated intractable disease for which no cure has yet been established. Vitiligo vulgaris is a disease that causes vitiligo throughout the body due to lack of melanocytes.

There are various theories about its etiology, but it is thought that the induction of an abnormal immune response to melanocytes is particularly important for the onset and progression of symptoms.

Around ten years ago, our laboratory started researching this area. Around five years later since then, it moved to the Institute of Technology, Akashi College from Osaka University. Looking back on the research and the crossroads of life, we are deeply moved by the diversity of life.

Lively and talented young people grow up in a fun and sometimes tough competitive manner. Your future is unknown, hopeful, and expected to be full of further discoveries. Looking back over the years, you can gain a better understanding of the diversity of life.

This book is written to introduce our attempt to deepen the understanding on melanogenesis, its diversity, from atomic nuclei and electrons world. At the beginning, Chap. 1 introduces an overview of melanin chemistry. Chapter 2 introduces theoretical works performed on dopachrome conversion. Chapter 3 introduces reactions of dopaquinone and *o*-quinones. At the end, Chap. 4 summarizes our findings and introduces future challenge.

Fukuoka, Japan Ryo Kishida
Akashi, Japan Susan Meñez Aspera
Akashi/Osaka, Japan Hideaki Kasai

Acknowledgments

R. Kishida acknowledges the Japan Society for the Promotion of Science (JSPS) for the financial support by Grant-in-Aid for JSPS Research Fellow (17J01276).

S. M. Aspera acknowledges the Ministry of Education, Culture, Sports, Science and Technology (MEXT) through their Quantum Engineering Design Course (QEDC) of Osaka University, Marubun Foundation and Kansai Research Foundation for their financial support. This work is supported in part by JST ACCEL grant number JPMJAC1501 "Creation of the Functional Materials on the Basis of the Inter-Element-Fusion Strategy and their Innovative Applications", MEXT Grant-in-Aid for Scientific Research (16K04876), and JST CREST Innovative Catalysts and Creation Technologies for the Utilization of Diverse Natural Carbon Resources: In-situ atomic characterization of catalytic reactions for the development of Innovative Catalysts (No. 17942262).

H. Kasai would like to thank the students and staff for the wonderful years they have shared with him.

The authors thank Professor Emeritus Shosuke Ito and Professor Emeritus Kazumasa Wakamatsu for their continuous support and constructive discussion with warm words and brilliant insight.

The authors would also like to thank Dr. Shin'ichi Koizumi of Springer Japan for his great support in preparing and writing this book.

Contents

Chapter 1
Melanin Chemistry

Abstract The history and the current status of melanin chemistry are introduced briefly. It is known now that melanin, a pigment found in animals, consists of two types of oligomeric unit: eumelanin (black to brown) and pheomelanin (yellow to reddish brown). The color of the skin, the hair, and the eyes is mainly determined by the ratio of eumelanin/pheomelanin production as well as the total amount of melanin. Among various reaction steps involved in the biosynthesis of melanin, there are two branched reactions that directly affect the composition of melanin: (i) reaction of dopachrome and (ii) reaction of dopaquinone. We introduce our approach to gain an understanding of these reactions from atomic nuclei and electrons world.

Keywords Computational materials design (CMD®) · Melanin · Eumelanin · Pheomelanin · Melanocyte · Tyrosinase · Raper-Mason pathway

1.1 Overview

1.1.1 Significance of Fundamental Studies

Exploring microscopic behaviors of biochemical reactions is one of the challenges in science. Living organisms have controlled chemical reactions by various proteins, which had been naturally selected through evolution. Chemistry of biomolecules including amino acids and proteins is one of the keys to understanding the mechanisms of biochemical reactions.

Due to involvement of complex internal motions of biomolecules, mechanistic studies have frequently encountered difficulties in describing the reactions. Furthermore, modeling of biochemical reactions often requires environmental factors that are usually less significant in organic chemistry. For instance, hydrogen bonding, proton exchange with water molecules, and catalytic interactions with co-existing metal ions may participate in the reaction as necessary processes. Moreover, biochemical reactions usually appear in multi-step and multi-branch cascades, conferring diversity and flexibility to the biochemical systems.

The above-mentioned complexity might contribute in part to realization of the various functions of organisms. It is one of the goals of several fields, such as tissue

R. Kishida et al., *Melanin Chemistry Explored by Quantum Mechanics*,
https://doi.org/10.1007/978-981-16-1315-9_1

engineering, drug design science, and biomaterials, to artificially reproduce and improve such functions of organisms. Toward realization of such reaction design, fundamental understanding of microscopic behaviors of biochemical reactions is necessary.

Recent development of computational science for physics and chemistry problems has enabled us to simulate the electronic states of biomolecules based on quantum theory. Computational materials design (CMD®), which guides material functionalization or reaction prediction without fitting to experimental data, is now a generally accepted framework for creation of intellectual properties. First principles calculation, which only requires the atomic numbers and the atomic configurations, is a basis of CMD® realizing the analysis and the prediction of nanoscale effects on the material structures and compositions. To see the wide effectiveness of first principles calculation methods and future possibilities, this book reviews CMD® applications to biochemical reactions as case studies.

1.1.2 Classification of Melanin

Among the natural pigments formed by living organisms, pigments called melanins with specific chemical properties and structures have been universally discovered. Melanin shows various protective functions (or conservatively speaking, energy-converting responses), including ultraviolet (UV) absorption and scavenging of reactive oxygen species (ROS). Although from the 19th century natural pigments exhibiting black to dark brown color were vaguely called melanins, the chemical characteristics of these pigments such as their composition and the structure were not clear. In the 20th century, research progress on these pigments significantly advanced the chemical understanding. In the stream of these chemical studies, the term melanin was defined and classified based on its chemical characteristics and became almost the same meaning as the one currently used [1].

From a chemical point of view, melanin can be classified into two types. One is brown to black eumelanin, and another is yellow to reddish brown pheomelanin [2]. Natural melanins are mixtures of these two types of melanin, and the mixture ratio forms the various colors of the animal body [3]. Eumelanin is synthesized from indollic monomers while pheomelanin includes benzothiazine and benzothiazole as monomers.

In the field of cell biology, melanin-synthesizing cells (pigment cells) have been studied up-to-date. In the case of homotherm, there are two types of pigment cell: melanocyte and retinal pigment epithelium (RPE) cell. Pigment cells contain specialized organelles for melanin production called melanosomes [4].

Melanocytes are neural crest-derived cells, which deliver melanin-containing melanosomes to the skin and the hair [5]. In the case of poikilotherm, instead of melanocytes in homotherm, neural crest-derived melanophores exist as melanin-containing cells that control the body colors by varying their distribution.

RPE cells are optic cup-derived cells, and form the pigmented layer in the development of the retina. Melanin pigments in the RPE cells are responsible for absorbing light. Unlike cutaneous melanins, these optic cup-originated melanins are actively formed only before the completion of the differentiation to the RPE cells, and then show slow turnover (i.e. a dynamic equilibrium between production and decomposition). While the skin and the hair colors depend on transportation of melanins from the melanocytes to the outer layer of the epidermis, the RPE cells directly use the intracellularly synthesized melanins for their light absorption [6]. Aside from melanins synthesized in pigment cells, there are also melanin-like pigments in the neurons located at the substantia nigra and the locus coeruleus of the brain. These pigments are called neuromelanins and their biological effects, such as binding of Parkinson's disease-related heavy metal ions and molecules (e.g. methylphenylpyridine: MPP+), have been pointed out [7, 8].

1.1.3 The Significance and the Scope of Melanin Study

Melanin study covers wide branches of science. From the biodiversity point of view, melanin-based pigmentation is an important biology topic as it contributes to the various visual phenotypes. Understanding the origin of the visual phenotypes requires connecting the genetics of the color polymorphism with knowledge of the body coloration. Dermatologists have been struggling to explore the etiology/pathogenesis and the prevention/treatment of melanin- or melanocyte-associated diseases including pigmentary disorders, such as oculocutaneous albinism (OCA) and vitiligo vulgaris, and the malignant tumor of melanocytes, namely melanoma.

As the main color-determining factors are the amount of melanin and its composition (the ratio of eumelanin/pheomelanin), the lesions in pigmentary disorders display an irregular amount of melanin production, melanin transportation, and number of melanocytes. OCA is a congenital disorder characterized by dysfunction of melanin production, and designated as an intractable disease in Japan. Vitiligo is an idiopathic leukoderma resulting from loss of melanocytes, which can cause a widespread depigmentation for non-segmental cases. Although the etiology of vitiligo is still controversial, irregularly sensitized immune responses are widely accepted as an important factor contributing to the initiation and the progression of vitiligo [9].

Aside from pigmentary disorders, controlling the biosynthesis and the transportation of cutaneous melanins has been a longtime theme in cosmetic sciences. To achieve this artificial control of pigmentation, a multi-disciplinary investigation of melanin formation, involving chemistry, biology, and dermatology, is necessary.

Melanoma is a notorious cancer which has poor prognoses. As surgical resection of metastatic and invasive melanoma is poorly effective, chemotherapeutic agents have been used for the treatment of advanced melanoma. Although alkylating agents have a long history of application to chemotherapy, they are still not satisfactorily successful because of their limited response rates (efficacies) and their severe adverse effects.

Recently developed molecularly targeted drugs (e.g. targeted to V600E/K missense mutation in the tyrosine kinase BRAF for inhibition of the cancer proliferation, and to the negative regulatory immune receptor PD-1 for recovery of the cytotoxic immunity) showed improved response rates [10, 11]. However, there are still unsolved puzzles like the acquired resistance [12] as well as the adverse effects. The incident rate of melanoma has shown an increasing tendency in recent years, indicating in part a relation with increased UV exposure due to ozone layer depletion [13]. Therefore, developing more effective anti-melanoma agents is an urgent issue.

Melanin is also attracting attention from condensed matter physics and material sciences, as well as chemistry, biology, and medical sciences. The protective light barrier function of melanin is due to the highly efficient energy conversion from the absorbed light to heat [14]. Such energy-converting properties of melanin have been investigated with a focus on their dynamics [15]. As well as the energy dissipation to heat, photoconductive properties, which convert light energy to electric current, was also observed [16]. From investigations on electric conduction through melanin, bi-stable conduction characteristics, which drastically switch its resistances at a certain threshold voltage, were observed [17–20]. As such, properties of melanin might be useful for electrical applications; several fabrication methods of melanin materials were reported especially in a thin film form [16]. Biocompatible and biodegradable nervous tissue responses of a fabricated melanin thin film were also confirmed, both in vitro and in vivo [21]. This indicates the feasibility of melanin films for scaffold applications in regeneration of nervous and muscular tissues. Furthermore, a coating method by synthetic melanin using dopamine (dopamine-melanin) was reported as a versatile surface-functionalization technique, which makes various combinations of organic and/or inorganic (and metallic) surfaces adhesive to each other, indicating feasibility of melanin also in applications to surface/interface sciences [22].

In summary, melanin study covers wide science fields, including materials science, chemistry, and biology: synthetic melanin materials (e.g. thin films and adhesives), the melanin biosynthesis system, and the reaction environment for the biosynthesis, namely melanocyte. The reason for this broad attention to the melanin synthesizing system might come from the connection to the visually simple phenotype “color”, the oxidatively active reaction environment, and the wide variety of responses to chemical and physical stimuli like the presence of ROS and light irradiation. Thus, various research topics could emerge from exploration of this specialized environment for melanin biosynthesis. Accordingly, it is a fundamental problem to ask mechanisms by which melanin biosynthesis proceeds and factors that can regulate melanin biosynthesis.

The enzymes controlling melanin biosynthesis has been identified together with their enzyme reactions. In spite of these existing enzymes, as described in the later chapters, most of the processes involved in melanin biosynthesis spontaneously take place without participation of enzymes. The presence of such spontaneous reactions emphasizes a nature of non-enzymatic control of melanogenesis. As such control of reactions does not exhibit very specific binding nature unlike typical enzymatic processes, understanding melanin biosynthesis requires approaching unexplored frontiers.

1.2 Analysis Techniques for Melanin Chemistry

The study on melanin has the very long history from the latter 19th century to the present. In the earlier phase of study, it was not trivial how melanin should be defined. The chemical structure and the composition of melanin depend on the synthetic condition. It is also difficult to completely analyze the chemical structure and the composition of melanin from conventional techniques [23]. The difficulty in analyzing melanin samples lies in the strong carbon–carbon covalent bonds, which connect the constituent monomer to each other, making the separation into monomers by physicochemical technique such as the chromatography almost impossible. Besides, the interpretation of the structure is difficult because such bondings could form not only in a direction creating linear structures but also the branching structures. In contrast, more loosely bonded linear structures are commonly found in many natural macromolecules, such as the glycosidic bonds in saccharides, the peptide bonds in proteins, and the phosphodiester bonds in nucleic acids. Furthermore, melanin is poorly soluble in most of solvents, making its analysis much more difficult.

Up-to-date, there is a huge amount of science literatures with a term "melanin" as a keyword. Thus, the interpretation of these must take the origin of the used melanin into consideration; melanin depends on, the species if taken from a living tissue, the synthetic methods if chemically synthesized in lab, and the passage or the culture condition if formed in cultured cells, for instance. Therefore, it is advisable to learn what is accessible by analysis techniques for melanin samples as a first step. This section reviews widely used analysis methods in melanin chemistry.

The great effort on clarifying melanin biosynthesis revealed the building monomers of melanin. Eumelanin is constructed from two monomers, 5,6-dihydroxyindole (DHI) and 5,6-dihydroxyindole-2-carboxylic acid (DHICA) (Fig. 1.1), while pheomelanin is constructed from 7-(2-amino-2-carboxyethyl)-5-hydroxy-2H-1,4-benzothiazine, 8-(2-amino-2-carboxyethyl)-5-hydroxy-2H-1,4-benzothiazine, their 3-carboxy derivatives, and 6-(2-amino-2-carboxyethyl)-4-hydroxy-2H-benzothiazole (Fig. 1.2).

These monomers are bonded via oxidative reactions, and then thought to form higher order structures. Although it is not possible to separate a melanin into these monomers as mentioned above, there are several chemical reactions that can be used to indirectly find the eumelanin/pheomelanin ratio in a mixed melanin and the DHI/DHICA ratio in a eumelanin. This is called the chemical or the oxidative

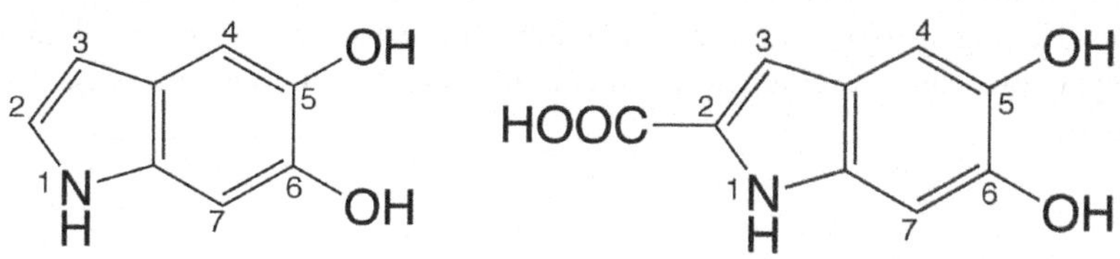

Fig. 1.1 Eumelanin building monomers. 5,6-Dihydroxyindole (DHI: Left) and 5,6-dihydroxyindole-2-carboxylic acid (DHICA: Right)

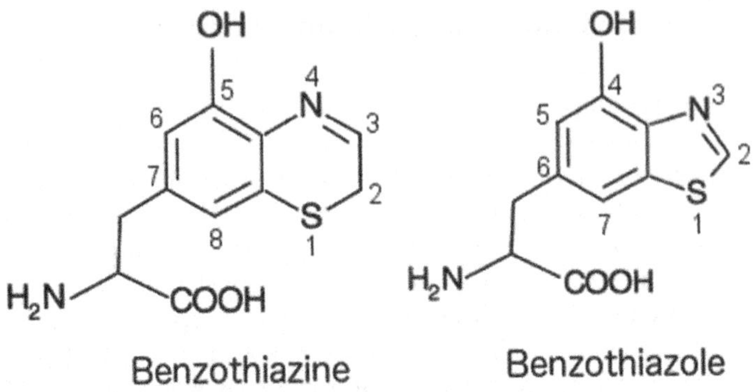

Fig. 1.2 Pheomelanin building monomers. 7-(2-amino-2-carboxyethyl)-5-hydroxy-2H-1,4-benzothiazine (Left), 6-(2-amino-2-carboxyethyl)-4-hydroxy-benzothiazole (Right). Carboxyl substitutions of 3H of benzothiazine and 2H of benzothiazole are also involved

degradation method, and was proposed by Ito et al. [24–26]. This degradation method originally used acidic potassium permanganate ($KMnO_4$) for oxidizing eumelanin, and hydroiodic acid (HI) for reductive hydrolysis of pheomelanin. This method requires complicated operations of experiment and the DHI/DHICA ratio was not available. Then, Ito et al. improved the original method by using alkaline hydrogen peroxide (H_2O_2) instead of $KMnO_4$ and HI.

The oxidative degradation by alkaline H_2O_2 results in the formation of four marker molecules (Fig. 1.3). One of the resulting markers is pyrrole-2,3,5-tricarboxylic acid (PTCA), which is a specific marker for DHICA-derived eumelanin. The amount of degradated DHI-derived eumelanin is also available by quantifying a specific marker pyrrole-2,3-dicarboxylic acid (PDCA), which was not quantifiable in the $KMnO_4$ oxidation. The pheomelanin amount are analyzed by markers for benzothiazole units, thiazole-2,4,5-dicarboxylic acid (TDCA), and thiazole-2,4,5-tricarboxylic acid (TTCA). The eumelanin/pheomelanin and the DHI/DHICA ratio can be calculated by analyzing the PTCA/TTCA and the PDCA/PTCA ratio, respectively. The degradation product containing the four marker molecules can be separately quantified by high performance liquid chromatography (HPLC) and UV absorption spectroscopy. This method is recognized as the standard for melanin compositional analysis, which has been used to find the relationship between the visual phenotype (color) and the genotype of various tissues with the aid of chemical understanding. The result of this chemical degradation can be said to define "chemical phenotype" of melanin samples. Having available the chemical phenotype as well as conventional the visual phenotype, one can now characterize melanin samples with much more quantitative information.

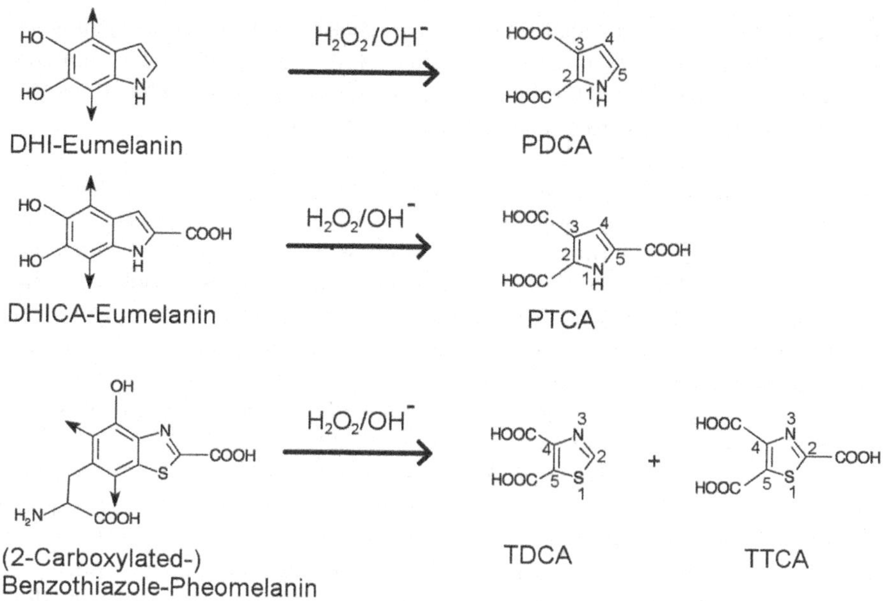

Fig. 1.3 Chemical degradation of melanin to form markers for melanin analysis. DHI- and DHICA-derived unit in eumelanin gives pyrrole-2,3-dicarboxylic acid (PDCA) and pyrrole-2,3,5-tricarboxylic acid (PTCA), respectively, as the degradation products by alkaline H_2O_2 oxidation. Benzothiazole unit in pheomelanin gives thiazole-4,5-dicarboxylic acid (TDCA) and thiazole-2,4,5-tricarboxylic acid (TTCA) as the degradation products by alkaline H_2O_2 oxidation. Note that the other products (not specific to melanins) are omitted for simplicity

1.3 Biosynthesis of Eumelanin—Formation of Dopaquinone and Dopachrome

1.3.1 Identification of Tyrosinase

This section reviews background on biosynthetic reactions to form the two pivotal intermediates, dopaquinone and dopachrome, which are located at "branching points" of the biosynthesis. These reactions are involved in the biosynthesis of eumelanin. Biosynthesis of melanin, namely melanogenesis, is a complex process via various unstable intermediates. Due to the short life of the intermediates, earlier studies were not able to capture some of the intermediates. The term melanin was probably first given by Berzelius in 1840 to refer black animal pigments [1]. The history of melanin chemistry was started with identification of enzymes participating melanogenesis.

In 1895, Bourquelot and Bertrand identified an enzyme tyrosinase (Tyr) in the extracts of mushrooms [27]. Tyrosinase catalyzes the oxidation of the substrate tyrosine. The formation of black pigments was confirmed by this tyrosinase-catalyzed reaction. This finding revealed the precursor tyrosine and the product melanin. The presence of tyrosinase was also subsequently reported in some of the plants, the insects, the fungi, and the marine organisms. Thus, the presence of mammalian tyrosinase was also expected [28]. However, identification of mammalian tyrosinase had been not straightforward to conclude for several years.

As an example, extracts of horse melanoma were able to convert tyrosine to melanin, whereas no tyrosinase activity was demonstrated by using fetal rabbit skin [29]. Bloch demonstrated important findings and tried to explain this puzzle, although not completely successful. Bloch immersed frozen sections of pigmented human skin in a solution of 3,4-dihydroxyphenylalanine (dopa), which is a hydroxylated tyrosine. The immersed tissues presented black pigments. This method is used even today to confirm the presence of melanocytes, and called the dopa staining. In contrast, immersing in tyrosine did not result in this pigment formation. From this result, Bloch proposed that "dopa-oxidase", which catalyzes the oxidation of dopa, is present in mammalian skins, whereas tyrosinase does not exist [30].

In this connection, it is noted that Raper isolated dopa in crystalline form, which was generated by oxidizing tyrosine with plant- or insect-derived tyrosinase [28]. The formed dopa could be further oxidized to form the final product melanin. Therefore, dopa is an intermediate of melanogenesis of, at least, the plants and the insects.

In 1942, Hogeboom and Adams demonstrated the oxidation of tyrosine and dopa when added to extracts of mouse melanoma, as confirmed by oxygen uptake. This oxidation was followed by black pigment formation, and thus the presence of tyrosinase as well as dopa-oxidase in mammalian tumors was indicated [30]. The tyrosinase and dopa-oxidase activity were separately obtained by fractionation using ammonium sulfates solution upon centrifugation of the melanoma.

Greenstein et al. also found tyrosinase and dopa-oxidase activity in the extracts of human melanoma [29]. The above studies on mammalian melanomas demonstrated lower tyrosinase activity than that of dopa-oxidase in general, and then tyrosinase activity could be hardly extractable depending on the tissues. Use of melanomas, which arose from sufficient quantities of melanocytes, is thought to be advantageous for finding the tyrosinase activity.

However, the hypothesis of co-existing tyrosinase and dopa-oxidase was denied after a while. In 1949, Lerner et al. reinvestigated tyrosinase activity from mouse melanoma [31]. The tyrosinase-catalyzed oxidation is initially very slow or lagged before the reaction begins, and then becomes fast later. This lag is referred to as the induction period. Lerner et al. found a shortened induction period by adding dopa, and formation of dopa by this catalyzed oxidation as an intermediate. It was also pointed out that dopa is readily oxidized even without enzymes above pH 7.0, indicating that the previous study by Bloch could overestimate the activity on dopa oxidation since the pH was at 7.4.

Lerner et al. emphasized the difficulties in separately evaluating the activity on tyrosine and dopa oxidation upon fractionation in the presence of the factor reducing the induction period. Based on this point, Lerner et al. proposed that tyrosinase and the previously hypothesized dopa-oxidase must be considered as the same enzyme, and thus the term tyrosinase should be only recommended.

1.3.2 Exploration of Raper-Mason Pathway

In parallel with the above-mentioned studies on tyrosinase, there was research progress on intermediates involved in melanogenesis. The oxidation of tyrosine to form an intermediate dopa is followed by the formation of another intermediate presenting red to orange color, which further spontaneously changes to colorless, and then converts into the final product black eumelanin. The colorless and the red to orange compounds were identified from a solution by Raper and Mason. Raper showed that the colorless compounds possess nitrogen atom(s) in the non-amino form [32]. The non-amino nitrogen was thought to be involved in a ring structure resulting from intramolecular cyclization (ring-closure reaction via bonding between amino N and benzene ring C). As tyrosine and dopa are not cyclizable, the oxidized substance of them, namely dopaquinone (a kind of *o*-quinones), must be responsible for the cyclization [33]. Raper successfully isolated the (*O*-methyl-substituted) colorless compound in crystalline form by employing several conditions to avoid autooxidation of the compounds, including careful control of atmosphere [33]. The CHN analyses and measurement of melting points were conducted with mixed synthetic compounds. As a result, DHI was identified as the major product, whereas DHICA was principally formed when the decolorization was conducted with sulfurous acid (an acidic reducing agent) [33]. Currently, these products were also identified by the HPLC retention time, and simplified preparation methods for DHI, DHICA, and *O*-methyl derivatives were established [34]. Raper also indicated that the red-to orange-colored pigment is probably 2,3-dihydroindole-5,6-quinone-2-carboxylic acid, namely dopachrome, based on the identified structure of the decolorization product DHI.

In 1948, Mason spectrophotometrically followed the tyrosinase-catalyzed oxidation of dopa to investigate the oxidation mechanism [35]. The oxidation of dopa at neutral pH gave rise to a sequential change in UV-visible spectra, in which at the first conversion the absorption maxima appeared at 305, 475 nm (red to orange), and next at 275, 298 nm (colorless), followed by 300, 540 nm (purple) of wavelength, and then finally the spectrum showed a broad absorption profile (black), corresponding to eumelanin. The absorption maxima at 275, 298 nm confirms the formation of DHI since its *O*-methyl derivative gives an identical absorption profile. At strongly acidic conditions (pH 1.3 or 2.0), instead, an absorption maximum at 310 nm was observed, which is identical to that of *O*-methylated DHICA. Although the purple pigment (called melanochrome) with absorption maxima at 300, 540 nm was thought as 5,6-indolequinone (IQ), subsequent reports denied this possibility and identified this pigment as a newly identified dimers [36]. The color of IQ was later found to be yellow [37]. The remaining absorption maxima at 305, 475 nm of the red to orange pigment were almost identical to adrenochrome and rubreserine (a kind of aminochromes), indicating that structurally related dopachrome is the corresponding compound.

The spectrophotometric findings given by the Mason's experiments provided a foundation for the development of melanin chemistry. Thus, the intermediates

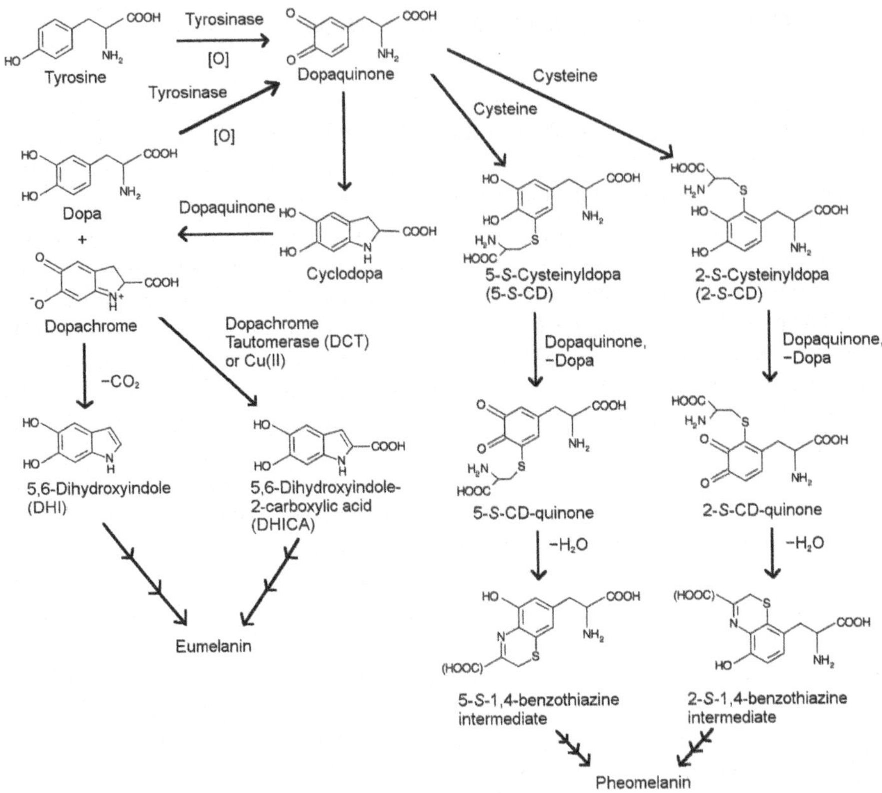

Fig. 1.4 Melanin biosynthesis pathway bearing the names of contributors to melanin chemistry. This reaction route is called Raper–Mason pathway

involved in the oxidative conversion of tyrosine to give eumelanin were almost entirely identified. The reaction route for the melanin biosynthesis is called Raper–Mason pathway, bearing the names of the above-introduced two contributors to melanin chemistry (Fig. 1.4). The term "melanocyte" was also proposed for the pigment cell responsible for melanin production around that time (1951) [38].

1.3.3 Identification of Dopachrome Tautomerase

Although it has been long believed that tyrosinase was the only enzyme involved in melanogenesis, in 1980, Körner and Pawelek isolated a melanocyte-specific enzyme which catalyzes the conversion of dopachrome into DHI and/or DHICA [39]. Körner and Pawelek extracted this enzyme from cultured mouse melanoma, and found the catalytic activity, which can be further enhanced by purification of the extracts. Based on this activity, the isolate was called dopachrome conversion factor (DCF).

Körner and Pawelek analyzed dopachrome consumption rate in the presence/absence of DCF by UV-visible spectroscopy, in which the product DHI was also detected with HPLC. The amount of decarboxylation (reaction to remove the carboxy group as carbon dioxide) was also quantified by ^{14}C labeling. As a result, it was revealed that DCF drastically accelerates dopachrome consumption and slightly accelerates decarboxylation, respectively. Production of DHI was detected regardless of the presence or absence of DCF (but faster in the presence of DCF). Considering that the acceleration of dopachrome consumption is more significant than that of decarboxylation, at that time DCF was thought to promote conversion of dopachrome to DHICA, and then DHICA spontaneously decarboxylates to form DHI.

In 1985, Körner and Pawelek identified the product of the DCF-catalyzed reaction as DHICA by spectrophotometric and ^{13}C NMR analysis [40]. At this time, it was recognized that DHICA was a transient intermediate and not a monomer. Thus, DHICA was thought to further convert to DHI via decarboxylation before it polymerizes to eumelanin. However, detailed analyses of eumelanins given by Ito et al. [41] revealed that DHICA-derived units were present in a content of nearly 50% in natural eumelanin [41]. It can be noted that the conducted analyses include the oxidative chemical degradation method, which is now a widely accepted method, as introduced in Sect. 1.2.

Integrating these results, the building monomers of eumelanin are regarded as DHI and DHICA. Briefly, DHI is produced in the absence of DCF catalytic action, while DHICA is formed if DCF acts on dopachrome. In this regard, it can be noted that there are also other (non-enzymatic) factors that catalyze dopachrome conversion into DHICA (see Chap. 3.).

Although some of previous studies have indicated that DHICA easily decarboxylates, this might be due to the use of mushroom tyrosinase; extracts of mushrooms may have effects of promoting decarboxylation from DHICA [42].

The product formed by the action of DCF is now identified as DHICA, and the conversion of dopachrome to DHICA is a protolytic tautomerization. Based on the understood action of this enzyme, a more specific new name "dopachrome tautomerase: DCT" was proposed, instead of DCF [42]. The name DCT is currently used for mammals, although DCF is exclusively used for insects. This is because, in the cases of insects melanogenesis, there is an enzyme that promotes dopachrome conversion to DHI but not to DHICA unlike mammals [43].

DCT has an amino acid sequence, which is very similar to tyrosinase, and belongs to a family called tyrosinase-related proteins (TRPs). (Therefore, DCT is also called TRP2.) I.J. Jackson et al. analyzed the amino acid sequence of mouse melanoma DCT. As a result, the sequence of DCT was found to be similar with tyrosinase and an enzyme called TRP1 expressed in melanocytes [44]. Analogous with tyrosinase, where copper ions at the active site are bound with three histidine residues, DCT was also expected to contain metal ions such as copper because histidine residues were located in similar places.

Although the possibility of copper or iron ions as DCT-containing metal ions has been predicted, experiments by F. Solano et al. revealed that the metal ions contained in DCT are zinc ions [45, 46]. Solano and colleagues analyzed the composition of metal ions by atomic absorption spectroscopy, and showed that DCT contained almost no copper or iron ions, but it contained zinc ions. Furthermore, after removing metal ions using cyanides (or other chelating agents) and then recombining with various metal ions, zinc ions showed the largest restoration of the enzymatic activity of DCT. Unlike copper ions, zinc ions hardly cause oxidations, indicating the markedly different function of DCT.

1.3.4 Reinvestigation of Tyrosinase Actions

The existence of the induction period of tyrosinase and its shortening by dopa have long been unresolved. This puzzle has gradually become resolved as melanogenesis has been investigated in more detail. Tyrosinase has long been known as a "copper protein" containing two copper ions, which are responsible for the oxidation reaction [27]. A study of the crystal structure of mushroom tyrosinase [47] showed the structure of the active site containing copper ions.

At the active site of tyrosinase, there is a pair of copper ions, each coordinated with three nitrogen atoms in histidine residues, and the pair center is able to bind oxygen and hydroxyl ions. The formation of a catechol (dopa) from a monophenol (tyrosine) in melanogenesis indicates that tyrosinase should have an ability to uptake and transfer oxygen (monooxygenase activity). Considering that tyrosinase-catalyzed oxidation consumes oxygen in the air [30], tyrosinase must be able to uptake oxygen from the air again after the reaction to restore the catalytic activity.

Most of the isolated tyrosinase is present in a state called *met*-tyrosinase form, in which Cu(II) ions are combined with a hydroxyl ion. This form cannot uptake oxygen any more because there are no oxidation states higher than Cu (II). However, when the coppers are reduced to Cu (I) by reducing agents to release the bound hydroxyl ion (*deoxy*-tyrosinase form), it becomes available for oxygen uptake in the peroxide state (*oxy*-tyrosinase form) from the air. Furthermore, as the fourth form, an irreversibly inactivated *deact*-tyrosinase form is generated when the *oxy*-tyrosinase acted on oxidation of catechols or resorcinols (Fig. 1.5) [48]. It has been hypothesized that there may be some connection between the mechanism to re-uptake oxygens and the mechanism by which dopa shortens the induction period of tyrosinase.

The shortening of the induction period can be explained by considering that dopa, resulting from tyrosine oxidation, acts as a reducing agent, thereby reducing the copper ions of tyrosinase so that it can react with oxygen again. To demonstrate this mechanism, the amount of oxygen uptake was investigated. When 4-hydroxyanisole was used as a substrate for tyrosinase, an equimolar O_2 consumption with respect to the substrate was observed. In contrast, when tyrosine was used as a substrate, the O_2 consumption was 1.5-fold higher than the amount of substrate [49]. This is because dopa, resulting from tyrosine oxidation, is further oxidized by tyrosinase to consume more oxygen. This extra O_2 consumption (0.5-fold amount of the substrate) can be regarded as a stoichiometry of 2 mol of dopa to 1 mol of O_2. This stoichiometric ratio

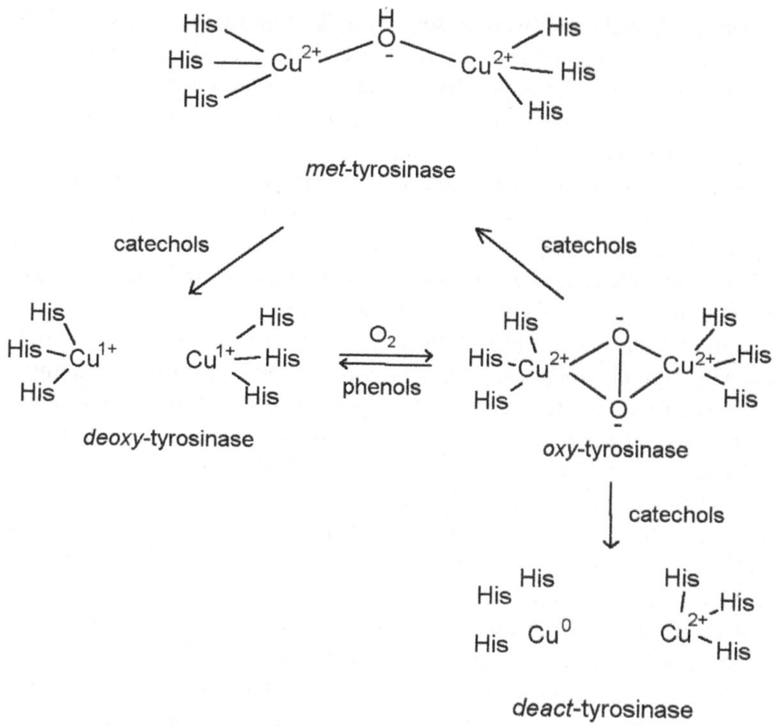

Fig. 1.5 Four isoforms of tyrosinase active sites. Cu ions display 3-fold coordinated structures with nitrogen atoms in histidine residues

indicates that both *met*- and *oxy*-tyrosinase acted on dopa oxidation at the 50% of probability. In other words, when *oxy*-tyrosinase meets dopa, tyrosinase is deprived of one of the two O atoms by dopa and changes into *met*-tyrosinase form, while when *met*-tyrosinase meets dopa, *deoxy*-tyrosinase form will be generated, where oxygen uptake will take place.

This model mechanism to explain the shortening of the induction period by the dopa redox reactions is not only the possibility, but could also be replaced by another mechanism, such as allosteric effects of dopa-tyrosinase (regulation of the enzyme activity by specific binding of an effector molecule at a certain site other than the active site). Nevertheless, having reported further supporting evidences [50], this model is currently a widely accepted mechanism. From earlier studies on melanin chemistry, it has been well known that tyrosinase-catalyzed oxidation of tyrosine results in the formation of dopa. Although whether or not dopa is a direct product of tyrosine oxidation, which had been a subject of controversy for a long time, an experimental evidence that supports indirect formation of dopa was presented [50]. *N*,*N*-dimethyltyramine and *N*,*N*,*N*-trimethyltyramine, which have structures very similar to that of tyrosine, were oxidized in the presence of tyrosinase, and the oxygen consumptions were measured. As a result, equimolar O_2 consumptions with respect to the substrate were observed, indicating that dopa-like catechol was not generated. The reason for choosing these molecules is to prevent the amino N from intramolecular cyclization by introducing protecting groups. Accordingly, if cyclization does not occur, the extra O_2 consumption (0.5-fold amount of the substrate) also does not occur. As described above, oxidation at O_2 /substrate = 1.5 would correspond to the formation of catechol such as dopa. Therefore, it should be interpreted that dopa was indirectly formed after the formation of dopaquinone. This indirect formation is likely to take place by reduction between uncyclized and cyclized dopaquinone (i.e. dopaquinone and cyclodopa), namely the redox exchange reaction, together with the formation of dopachrome.

Dopaquinone has a UV absorption at a wavelength maximum around 380 nm [51, 52]. Although chemical properties of dopaquinone and similar *o*-quinones had been discussed for a long time as mentioned above, direct observations had not been reported because of the too short lifetime. Tyrosinase-catalyzed oxidation was too slow as compared to cyclization of dopaquinone and the redox exchange. Therefore, it was not possible to spectroscopically identify dopaquinone because the production rate was lower than the consumption rate. This problem has become resolved by pulse radiolysis techniques, which were employed to track the production and the consumption of dopaquinone.

In pulse radiolysis, H_2O molecules are ionized and dissociated to form $^{\cdot}OH$ radicals in KBr or NaN_3 solutions (saturated with N_2O) by irradiating a high-energy pulse. By the chain reactions, the formed $^{\cdot}OH$ radicals further ionizes and converts Br^- or N_3^- into $Br_2^{\cdot}$ or $N_3^{\cdot}$, which are capable of one-electron oxidation. The one-electron oxidation reaction between $Br_2^{\cdot}$ or $N_3^{\cdot}$ and tyrosine is fast enough to observe the absorption peak near 380 nm, which is responsible for the formation of dopaquinone (Fig. 1.6). This was first shown in an experiment by Chedekel et al. in 1984 [53].

With the development of analytical methods, most of the processes in melanogenesis have become accessible to experiments. This section reviewed experimental studies that have been conducted to elucidate the processes of eumelanin monomer

$$H_2O \longrightarrow e^-_{aq} + {}^{\bullet}OH + H^+$$

$$e^-_{aq} + N_2O + H_2O \longrightarrow {}^{\bullet}OH + OH^- + N_2$$

$${}^{\bullet}OH + N_3^- \longrightarrow OH^- + N_3^{\bullet}$$

$$N_3^{\bullet} + QH_2 \longrightarrow N_3^- + QH^{\bullet} + H^+$$

$$2QH^{\bullet} \longrightarrow Q + QH_2$$

Fig. 1.6 Pulse radiolysis reactions in N_2O-saturated NaN_3 solutions. QH_2, $QH^{\cdot}$, and Q represent catechols (e.g. dopa), semiquinones (where one of H in catecholic OH is missing), and quinones (e.g. dopaquinone), respectively

formation. Melanin biosynthesis begins with tyrosine. Tyrosinase expressed in melanocytes catalyzes the oxidation of tyrosine to produce dopaquinone. When dopaquinone undergoes intramolecular cyclization, eumelanin production (eumelanogenesis) will take place. Cyclodopa formed by dopaquinone cyclization reacts with uncyclized dopaquinone to form dopachrome and dopa by means of redox exchange. Dopachrome can be spontaneously converted to DHI, but the presence of factors such as DCT promotes the conversion to DHICA. DHI and DHICA are crosslinked to each other via IQ and its 2-carboxyl derivative by a further oxidation reaction to form eumelanin. Figure 1.4 summarizes the melanogenesis route described here (and also the pheomelanin synthesis pathway).

1.4 Biosynthesis of Eumelanin—Oxidative Polymerization to Form Eumelanin

1.4.1 Identification of Oligomeric Molecules by Inter-Monomer Coupling

This section reviews the major findings that have been obtained so far regarding the processes of eumelanin formation by oxidative polymerization of DHI and DHICA. This oxidative polymerization process consists of two processes: catechol oxidation, which converts DHI and DHICA to the corresponding *o*-quinones (i.e. indolequinone IQ and its 2-carboxylated derivative IQ-CA, respectively), and inter-monomer coupling, in which the formed IQ (or IQ-CA) and the remaining DHI (or DHICA) are crosslinked by a carbon–carbon covalent bond (Fig. 1.7).

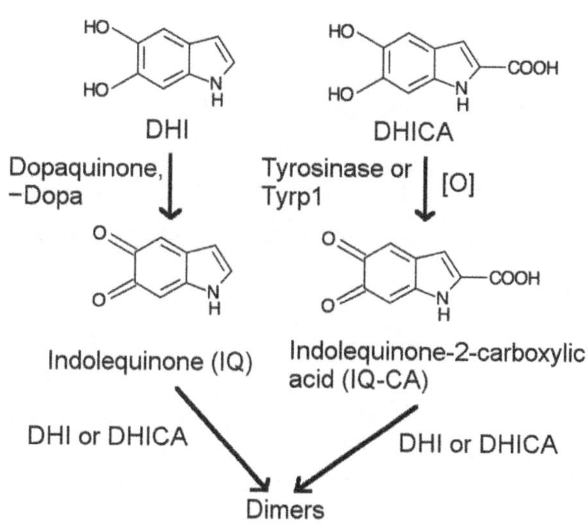

Fig. 1.7 Oxidative polymerization of eumelanin monomers. Catecholic oxidation of DHI and DHICA, respectively, gives indolequinone (IQ) and its 2-carboxylate derivative (IQ-CA), and the inter-monomer coupling gives corresponding dimers

As introduced above, it has been reported that dopaquinone produced upstream of melanogenesis can be an oxidizing agent for cyclic catechols, as a result of redox exchange between dopaquinone and dopa [54]. Then, dopaquinone, which is generated by the oxidation of tyrosine, can also oxidize catechols like DHI and DHICA, as well as cyclodopa. However, in the case of DHICA, catechol oxidation by dopaquinone is very slow. Thus, it has been pointed out that it is unlikely to be a process involved in actual melanogenesis [54]. At least in the case of mice, Trp1 (see Sect. 1.3.3; note that, unlike human proteins, only the first letter is capitalized for non-human-derived proteins) is able to catalyze the DHICA oxidation [55, 56]. Furthermore, human tyrosinase was shown to catalyze DHICA oxidation [57]. More recently, Cu (II) was reported to promote catechol oxidation of DHI and DHICA [58].

The inter-monomer coupling produces various structures of dimer: coupling between DHI and IQ at 2,4,7-carbons, and between DHICA and IQ-CA at 3,4,7-carbons (Fig. 1.8) [59–64]. Trimers and tetramers linked at 2,3,4,7-carbons have also been reported [62, 65–67]. Although the inter-monomer coupling between two 2-carbons does not usually occur, this 2-2' bonding can take place in the presence of Zn (II), Ni (II), and Cu (II) ions [59, 64, 68]. A heterodimer of DHI and DHICA with 2–4' linkage was also found [69]. Although no 3-carbon-linked DHI-IQ dimers have been identified, DHICA and IQ-CA are able to form 3–4' and 3–7' bonding. This crosslinking at 3-carbon becomes also possible when 2-carbon has methyl group as well as carboxyl group [70]. Crosslinking at 3-carbon was also reported by aging of eumelanin. The chemical degradation analysis (see Sect. 1.2.) of melanins in the fossilized ink sac from cuttlefishes in the Jurassic period detected considerable amounts of pyrrole-2,3,4,5-tetracarboxylic acid (PTeCA), which is usually less significant in usual melanin samples [71]. PTeCA has the fully carboxylated structure in its pyrolic ring, suggesting that a carbon–carbon covalent bond also exists at 3-carbon of DHI or DHICA. This aging was mimicked by heating eumelanins for

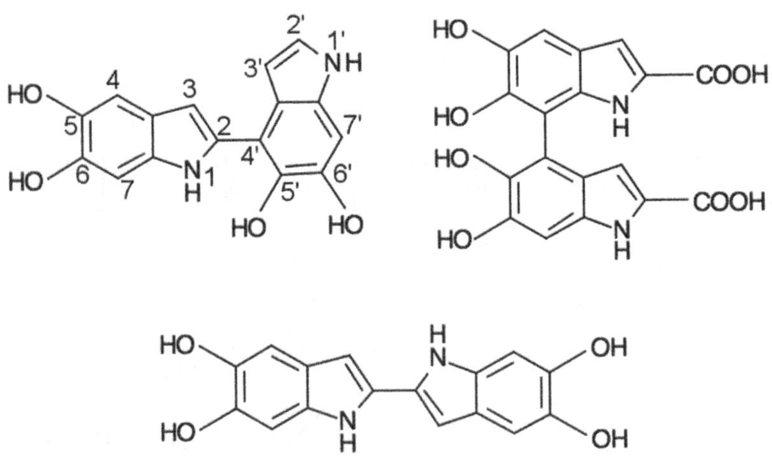

Fig. 1.8 Examples of dimers formed by the inter-monomer coupling in eumelanin biosynthesis

several days (100 °C for 18 days and 40 °C for 180 days), and chemical degradation produced increased amounts of PTeCA, reproducing the aging [72].

1.4.2 Theoretical Study of Monomer Polymerization and Melanin Structure Model

The above introduced experiments have successfully identified up to tetramers of DHI or DHICA homo-oligomers, and dimers of DHI-DHICA hetero-oligomers. However, the extent to which these coupling reactions proceed and how the reactivity changes by the polymerization remains unclear at present. For this reason, studies have been made on approaches based on theoretical calculations for the structures and properties of various oligomers composed of DHI or DHICA.

In early theoretical studies, the eumelanin structure was usually modeled by macromolecules, which were simply built by means of polymerized DHI and/or IQ (through coupling with IQ and/or DHI, respectively). Based on this model, H.C. Longuet-Higgins proposed that a mainly DHI-derived eumelanin should behave like a p-type semiconductor, while a mainly IQ-derived eumelanin is likely to behave as an n-type semiconductor [73].

Since both DHI and IQ have closed-shell electron configurations, the band theory will not find the presence of conduction carriers (electrons or holes) in the lower excited states, if only DHI and IQ constructed eumelanin in a straightforward manner. Experiments using electron spin resonance (ESR) have revealed that melanin exhibits paramagnetism, indicating that some of the monomer units of melanin have unpaired electrons [74]. Since semiquinone (SQ) obtained by one-electron oxidation of DHI has an unpaired electron, conduction carriers (electrons) can be generated when SQs are partially contained in eumelanin to create donor levels. Similarly, partially SQ-substituted IQ polymers, which arises from one-electron reduction, would have conducting holes with acceptor levels.

Pullman and Pullman employed Hückel approximation (or tight-binding approximation) to calculate the molecular orbital of the dimer linked at 3–7' carbons of two IQs [75]. As a result, the lowest unoccupied molecular orbitals (LUMOs) of IQ overlapped to each other to create a bonding orbital, yielding a decreased energy gap between the highest occupied orbital (HOMO) and LUMO (HOMO–LUMO gap). It was then considered that the band gap of eumelanin could be smaller when IQ content was increased with respect to the other monomer units (DHI and SQ).

However, a study of the band calculation for a one-dimensional polymer, in which IQs were periodically linked at the same binding sites, showed that the band gap became larger than that of the dimer [76]. Furthermore, the band gap of the one-dimensional polymer of SQ (at least in the case of 3–7' bonded structure) is smaller than that of IQ, and the one-dimensional polymer of DHI has a higher band gap than that of IQ with smaller band dispersion [77].

As described above, it has been long considered that eumelanin is a large extended polymer. However, X-ray diffraction study using synchrotron radiation rather suggested a stacked oligomer model as the melanin structure, in which about 4–5 monomers are bonded in a plane, and these protomolecules are stacked by π-π stacking in about four layers. [78, 79]. This model was later supported by observation using scanning tunneling microscopy (STM) [80]. Thus, the "chemical disorder model", which regards melanins as amorphous materials consisting of various oligomers, has become gradually accepted [81]. However, it should also be noted that the proposed planar oligomers have not yet been isolated in experiments, and only one-dimensional oligomers have been obtained as described above.

For DHI and IQ, first-principles calculations based on density functional theory (DFT) were performed [82, 83], and the electronic excitation energies were also calculated [84, 85]. Result confirmed that IQ has the electronic excitation energy lower than that of DHI. First-principles calculations, which were performed for DHICA and its oxidants (IQ and SQ), revealed that the presence of 2-carboxyl group significantly affects the stability and the HOMO–LUMO gap of the various oxidation forms [86]. In comparison with DHI-derived eumelanins, which are known to have considerable amounts of radicals, this study indicated that DHICA-derived eumelanins should contain negligibly small amounts of radicals. The electronic structure of DHICA dimer was also calculated, showing a red-shifted electronic excitation energy with respect to that of monomer [87]. These studies suggest that the HOMO–LUMO gaps of eumelanin monomers and their oligomers are highly dependent on their structures. This qualitatively explains the broad UV-visible absorption profile of eumelanin [81, 87].

Okuda et al. calculated the reactivity of DHI and its dimer using an index called the general-purpose reactivity indicator [88, 89] devised by Anderson and others [90]. This takes into account both effects of the electrostatic interaction and the charge transferring interaction. This is calculated using the atomic charges and the condensed Fukui function, the latter of which is the derivative of the atomic charge with respect to the total electron number with a negative sign [90–92]. The calculated reactivity results are consistent with the experimentally identified oligomers (dimers, trimers, and tetramers). For example, 2-carbon in DHI was shown to be highly reactive. Easier formation of tetramers was also indicated as compared to formation of trimers. The reactivity indicators showed that these reactions are mainly predominated by interactions involving charge transfer.

In this section, the major studies on the oxidative polymerization processes and the structural model of eumelanin were reviewed. Eumelanogenesis is a very complicated process, and its analysis is also difficult. There are still points that have not been clarified yet. However, by continuous efforts from both experimental and theoretical approaches, chemistry of eumelanogenesis is getting established as compared to pheomelanin production (pheomelanogenesis).

1.5 Biosynthesis of Pheomelanin—Reaction Process After Binding to Cysteine

Dopaquinone can react with thiols (R-SH) such as intracellular cysteine (Fig. 1.4). The binding of thiols takes place by nucleophilic addition with cysteine thiolate ion at 5-carbon and 2-carbon of dopaquinone, resulting in the formation of 5-*S*-cysteinyldopa and 2-*S*-cysteinyldopa, respectively [93]. (Both substituted dicysteinyldopa is also formed.) When these cysteinyldopas were oxidized by pulse radiolysis, the absorption maximum at around 310 nm of cysteinyldopas (note that the peak slightly shifts depending on the sulfur binding site) was immediately diminished, and then a new peak near 380 nm appeared instead. This new peak was further spontaneously replaced by two peaks around 330 and 540 nm [94]. Since the absorption maximum near 380 nm is a characteristic property of *o*-quinones, the oxidation of cysteinyldopas was likely to produce the corresponding *o*-quinones, namely cysteinyldopaquinones, as the initial products. The consumption rate of cysteinyldopas was proportional to the concentration of cysteinyldopas and of the molecules with the absorption maximum at 380 nm (considered to be cysteinyldopaquinones) [94]. In addition, the transient species with the absorption maximum at 380 nm were consumed at a rate proportional to their own concentrations, thereby producing a molecule having absorption maxima near 330 and 540 nm [94]. Here, the peak near 330 nm was stable for several tens of seconds, but the peak near 540 nm was unstable, which decayed by the first-order kinetics.

From the above spectrophotometric observation, a reaction scheme as shown in Fig. 1.9 was proposed. First, cysteinyldopa is oxidatively transformed to cysteinyldopaquinone, and the amino N in cysteine forms a bond with the carbonyl C in *o*-quinone to form a ring structure. (The carbonyl O is eliminated as H_2O together with the amino H.) The quinoneimine body thus formed is likely to be responsible for the absorption maximum at 540 nm. Through decarboxylation or tautomerization, this quinoneimine is converted to 1,4-benzothiazine (absorption maximum at 330 nm), which is a pheomelanin monomer.

The conversion to the quinoneimine was later demonstrated in a more direct manner [95]. 3,4-Dihydro-1,4-benzothiazine-3-carboxylic acid (DHBTCA), which is a reduced form of the quinoneimine, was subjected to pulse radiolysis. As a result, the radiolytic oxidation product showed an absorption peak at 540 nm, which is the same observed in the oxidation of cysteinyldopa, confirming the production of the quinoneimine intermediate [95].

The second-order kinetics of cysteinyldopa consumption can be explained by considering (partial) reduction of the quinoneimine by the unreacted cysteinyldopa, which produces DHBTCA (and cysteinyldopaquinone). This was proposed based on an HPLC analysis that showed the formation of DHBTCA [96]. This indicates that the quinoneimine and DHBTCA are in a chemical equilibrium. The formation of 1,4-benzothiazine was later directly confirmed by HPLC analysis of the $NaBH_4$ (or $NaBD_4$) reduction products [97]. Note that two types of 1,4-benzothiazine (i.e. 3-decarboxylated and carboxyl-retained cases) were found. Although HPLC cannot

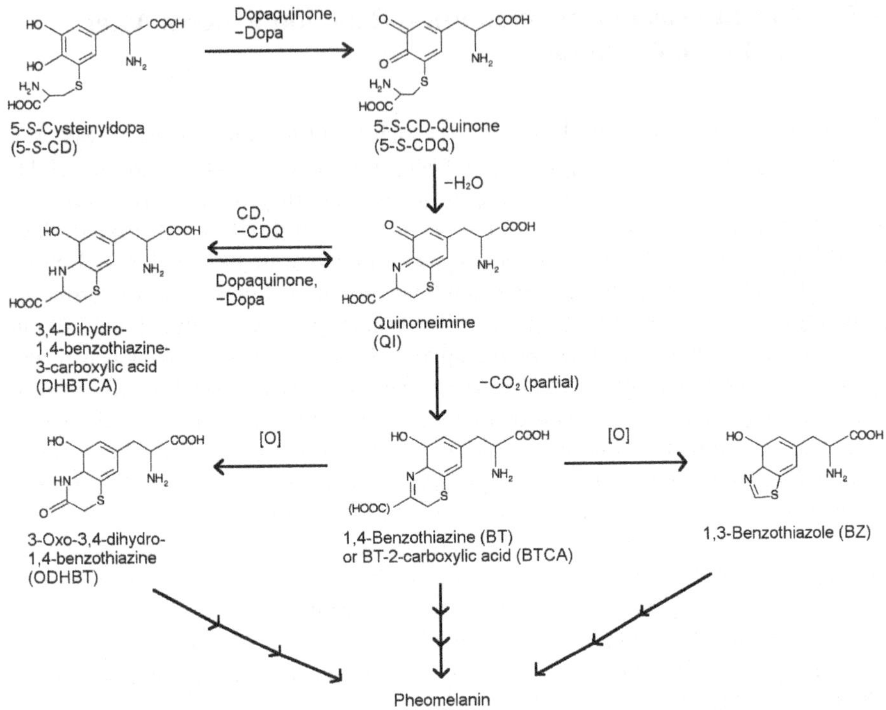

Fig. 1.9 Formation of pheomelanin building monomers (1,4-benzothiazine, 1,3-benzothiazole, and 3-oxo-3,4-dihydro-1,4-benzothiazine)

distinguish the $NaBH_4$ reduction product of 1,4-benzothiazine-3-carboxylic acid and DHBTCA, the use of $NaBD_4$ enabled selective detection of 1,4-benzothiazine-3-carboxylic acid.

Pheomelanin shows an absorption spectrum with a peak around 300 nm that attenuates toward longer wavelengths [98]. Although this peak position is slightly different from the absorption maximum at 330 nm of benzothiazine, possible subsequent products, namely benzothiazole and 3-oxo-3,4-dihydro-1,4-benzothiazine (3-oxo-3,4-dihydro-1,4-benzothiazine: ODHBT), have an absorption maximum near 300 nm. Therefore, pheomelanin was considered to also include benzothiazole and ODHBT as building monomers as well as benzothiazine [98]. This was confirmed by detailed analyses of tyrosinase-catalyzed oxidation of dopa in the presence of cysteine [99]. This reaction resulted in accumulation of an intermediate DHBTCA until the middle stage of the reaction, and then the further redox reactions gave rise to benzothiazine-based pheomelanin, which showed gradual degradation into benzothiazole-based pheomelanin. The formation of small amount of ODHBT was also demonstrated by HPLC analyses of the reaction mixtures.

In principle, tyrosinase may also be involved in the oxidation of cysteinyldopa. Nevertheless, tyrosinase activity is usually correlated with eumelanogenesis rather than pheomelanogenesis [100]. Therefore, tyrosinase-catalyzed oxidation of cysteinyldopa is not considered as a main factor of pheomelanin production. The evidence of the redox exchange reaction between dopaquinone and cysteinyldopa was given by a pulse radiolysis experiment [100]. Pulse radiolysis of dopa with added 5-*S*-cysteinyldopa resulted in a unique transient profile of light absorption, which is markedly different from the case without cysteinyldopa [100]. Although the absorption at 380 nm corresponds to both *o*-quinones (i.e. dopaquinone and 5-*S*-cysteinyldopaquinone), the molar absorption coefficient of 5-*S*-cysteinyldopaquinone is large enough for analyzing the redox exchange reaction, which simultaneously forms the two quinones. After the immediate formation of dopaquinone by pulse radiolysis, pseudo-first-order growth of 5-*S*-cysteinyldopaquinone was recorded, demonstrating the redox exchange reaction between dopaquinone and cysteinyldopa. In a similar manner, DHBTCA is oxidized by dopaquinone in pheomelanogenesis [99]. Thus, dopaquinone is an important oxidant in pheomelanogenesis.

Some of reported cysteinyldopa oxidation study were conducted with metal ions. Aside from the possible oxidizing functions of metal ions, it has been pointed out that metal ions may also significantly affect pheomelanogenesis as non-oxidative catalysts. For instance, Zn(II) suppresses the decarboxylation during the conversion of the quinoneimine into 1,4-benzothiazine [101]. In contrast, Cu(II) and Fe(III) promotes the decarboxylative pathway [102]. Furthermore, in the presence of Fe(III), increased benzothiazole-moiety in pheomelanin was obtained [102]. These metal ions are relatively rich in melanosomes [103, 104]. Therefore, the above findings are an important indication of physiologically relevant effects of metal ions on pheomelanogenesis.

This section reviewed the findings that have been revealed regarding pheomelanogenesis. These studies have established the overall picture of the reaction as shown in Fig. 1.9. Chapter 3 of this book focuses on the mechanisms of the earliest process, namely the formation of cysteinyldopa.

1.6 Melanin Chemistry in Relation to Melanocyte-Specific Cytotoxicity

1.6.1 Cytotoxic Effects of p*-Substituted Phenols on Melanocytes*

Due to the relatively low substrate specificity of tyrosinase in melanocytes, not only the melanogenic starting substances, namely tyrosine and dopa, but also structurally similar phenols and catechols are recognized by tyrosinase. This causes formation of dopaquinone-like *o*-quinones, resulting in melanogenesis-like reactions.

Specifically, 4-*tert*-butylphenol (4-TBP), monobenzone, 4-*S*-cysteaminylphenol (4-*S*-CAP), and rhododendrol (RD) are oxidized in the presence of mushroom tyrosinase [105–108]. Human tyrosinase-catalyzed oxidation of 4-TBP and RD was also reported [105, 109]. Thus, the formed *o*-quinones are usually highly reactive toward nucleophiles (chemical species that form a coordinate bond by donating an electron pair), and thereby the inherent chemical instability may influence on cytotoxicity.

The above phenols, at a certain level of high concentration, cause melanocyte-specific cytotoxicity as well as melanogenesis inhibition. Due to such cytotoxic effects, these phenols can be a double-edged sword to the use in dermatology and cosmetics; they may cause leukoderma as observed in pigmentary disorders as well as skin-whitening.

Cytotoxic effects of these phenolic compounds and its applications in dermatology have been widely investigated. 4-TBP and monobenzone are well-known compounds because of their depigmentation effects. Especially, monobenzone is now clinically used for depigmentation therapy, in which the patchy depigmented lesions of vitiligo patients are treated with monobenzone to make them uniformly white [110].

An in vitro study using melanocytes showed that 4-TBP and monobenzone activated oxidative response and unfolded protein response (UPR), and thus upregulated expression of pro-inflammatory cytokines [111]. These results indicate that 4-TBP and monobenzone may activate an autoimmune response through endoplasmic reticulum stress due to ROS generation. As a cytotoxic effect, 4-TBP induces apoptosis (a form of programmed cell death mainly through caspase activation pathways) [112], while monobenzone induces necrosis (an unregulated form of cell death/damage that does not follow apoptosis-associated signaling pathways) [110]. Another experiment using melanoma showed several immunogenic effects upon monobenzone treatment [113]. In the treated cells, melanosomes were subjected to autophagy (a mechanism by which cellular materials are degraded through membrane trafficking processes), tyrosinase was ubiquitinated (binding of ubiquitin to be degraded by proteases), and exosomes (membrane-bound extracellular vesicles) that contains melanoma antigen recognized by T-cell 1 (MART-1) and tyrosinase were released. A co-culture experiment of melanoma with dendritic cells showed activation of dendritic cells, and further addition of blood cells induced cytotoxic melanoma-reactive T-cells. These results suggest possible mechanisms of monobenzone-induced immunogenicity; monobenzone binds tyrosinase or other proteins to be an antigen, which is partly secreted outside and then (through cross-presentation by the activated dendritic cells) cytotoxic T-cells emerge.

4-*S*-CAP and its derivatives have been investigated as candidates of anti-melanoma drugs. An in vitro experiment showed that *N*-propionyl-4-*S*-CAP caused apoptotic cell death of melanoma and increased ROS production [114]. Furthermore, in a mouse melanoma model experiment, intratumoral injection of *N*-propionyl-4-*S*-CAP suppressed tumor growth and mediated cytotoxic TRP2-specific T-cells [114].

RD had been used in cosmetics as a skin-whitening agent in Japan until 2013, when it was recalled because of cytotoxic adverse effects causing vitiligo-like leukoderma. From an in vitro experiment, the viability of RD-treated melanocytes decreased in a dose-dependent manner, which could be associated with the observed UPR activation and be caused by induced apoptotic pathways [115]. From a chemical experiment, the formation of super oxide radicals was also shown when RD was oxidated with tyrosinase [108].

1.6.2 o-Quinones and Melanocyte-Specific Cytotoxicity

The above introduced cytotoxicity, in part, is presumably associated with the reaction between intracellular thiols and *o*-quinones generated by phenol or catechol oxidation. (However, note that there is also a report showing that the cytotoxicity of 4-TBP is not correlated with tyrosinase activity [112].) As previously stated, *o*-quinones are highly reactive toward nucleophiles. For example, *o*-quinones can bind intracellular thiols such as cysteine and glutathione (GSH), and proteins having cysteine residues (protein thiols). Since GSH acts as an antioxidant (e.g. GSH reacts with H_2O_2 to convert it to H_2O), depletion of GSH by binding with *o*-quinones would increase intracellular oxidative stress. Thus increased stress may stimulate endoplasmic reticulum stress, leading to cell death through apoptotic and/or other pathways. In addition, an elevated H_2O_2 concentration may induce up-regulation of tyrosinase activity [116], facilitating the production of *o*-quinones to cause cellular stresses in an accelerated manner. It has also been reported that the *o*-quinone generated by RD oxidation, namely RD-quinone, forms a pheomelanin-like pigment as the final product, through the binding of cysteine [117]. Pheomelanin is basically recognized as a pro-oxidant, which triggers ROS formation through photo-excitation and then affects cellular oxidative stress [118–120].

As described above, the importance of immune response has been emphasized, as well as cell injury caused by oxidative stress. From the viewpoint of the immune system stimulation, the binding properties of *o*-quinones with protein thiols would be important. As a unified explanation of the vitiligo mechanism, "Haptenation theory" has been proposed [121]. This focuses on the fact that the generated *o*-quinone is recognized as an antigen by binding with proteins, and that it induces cellular immune responses. Although small chemical species alone cannot elicit immune responses, some of them can form protein-bound complexes which are recognized by immune cells. Such chemical species are called haptens. After binding with proteins, as a possible mechanism to initiate immune responses, the complex may be ubiquitinated to be degraded by proteasome and/or engulfed by autophagy. The peptide fragments degraded here may be presented on the cellular surface with major histocompatibility complex (MHC) class I and II (Note that melanocytes also express MHC class II, as well as class I like antigen-presenting cells.), or be secreted by releasing exosomes, which activate antigen-presenting cells like dendritic cells. Through antigen presentation, activation of immune cells including CD8+ T-cells may occur, thereby melanocyte-specific cytotoxic T-cells will emerge and proliferate. Besides haptenation-associated immune sensitization, secretion of pro-inflammatory cytokines IL-6, which is triggered by UPR activation as in 4-TBP, monobenzone, or

RD treatment [111], may also be important for the progression of vitiligo lesions by inhibiting the regulatory T-cell functions [122].

Thus, these *o*-quinones have attracted attention as a selective cytotoxic drug. As mentioned above, this cytotoxicity is being considered for applications to depigmentation therapy (by monobenzone) and anti-melanoma treatments (*N*-propionyl-4-*S*-CAP, etc.). On the other hand, as in the case of RD-containing skin-whitening agents, this cytotoxicity acts in an unintended manner, causing adverse effects. To solve such melanin chemistry-related clinical problems, it is necessary to clarify the relation between *o*-quinone reactivity and cytotoxicity. However, our mechanistic understanding of *o*-quinone reactions is still far from complete due to the short lifetime of the participating molecules. Understanding melanin chemistry at the atomic and the electronic scale is an important step for predicting *o*-quinone reactivity.

1.7 Summary of This Chapter and Scope of This Book

Melanin is a mixed pigment of eumelanin and pheomelanin. Eumelanin is formed in the oxidative polymerization of DHI and DHICA, while pheomelanin is mainly built from benzothiazine and benzothiazole.

Section 1.2 of this chapter introduced that the ratio of eumelanin/pheomelanin and of DHI/DHICA are quantifiable and define the chemical composition of melanin. In Sect. 1.3, the initial processes of melanogenesis were reviewed. Tyrosinase-catalyzed oxidation of tyrosine or dopa produces dopaquinone, which then converts into DHI and DHICA via dopachrome, resulting in eumelanin production. This spontaneous conversion of dopaquinone is triggered by intramolecular cyclization. As mentioned in Sect. 1.5, in contrast, the binding of cysteine with dopaquinone causes pheomelanin production. As implications of melanin chemistry to dermatology and cosmetic science, Sect. 1.6 reviewed melanocyto-specific cytotoxicity of tyrosine/dopa analogs, which are oxidized by tyrosinase to cause melanogenesis-like reactions by forming reactive *o*-quinones.

If these melanogenesis-like reactions are manipulable by molecular design, more broadened applications of melanin chemistry to various fields are expected. Specifically, by making use of the photoelectronic properties, tissue compatibility, and biodegradability, melanin chemistry may also be applied to electronics and tissue engineering, as described in Sect. 1.1. In addition, in order to predict any adverse effects of drugs administered to melanocyte-containing tissues, it is necessary to understand the relationship between melanogenesis-like reaction and the cytotoxicity as described in Sect. 1.6.

Melanogenesis includes a branched reaction in the course of reaction. For example, dopaquinone is located at the branch point, and thus competitive processes, namely cyclization and binding of thiols, respectively produce eumelanin and pheomelanin. Furthermore, after dopaquinone cyclization, a eumelanogenic intermediate dopachrome undergoes spontaneous conversion to form the two possible eumelanin monomers DHI and DHICA. At this second branch point, DHI is formed when decarboxylation from dopachrome occurred, while the formation of DHICA proceeds by tautomerization with the retained carboxyl group. These branched reactions determine the chemical composition of melanin, namely the ratio of eumelanin/pheomelanin and the ratio of DHI/DHICA of eumelanin. So far, factors that influence these competitive reactions at the branch points have been investigated, although the mechanisms by which the branching occurs have not been clarified yet. Furthermore, chemistry of melanogenesis-like reactions initiated by tyrosine/dopa analogs is still an unexplored region. Therefore, it is a great challenge to establish a theoretical basis for understanding the whole picture of melanogenesis-like reactions that enables us to predict the reactivity for various cases of molecules.

In this book, we focus on two types of reactions in melanogenesis, which are thought to determine the properties of melanin. One is dopachrome conversion, which is responsible for the production of two types of monomers that build eumelanin. The other is dopaquinone conversion, which is responsible for eumelanin or pheomelanin production. The main objective is to give an explanation from a general point of view of how these reactions proceed at the atomic level as well as discussing chemical factors affecting the reaction pathway branching. Furthermore, we also aimed to understand melanogenesis-like reactions based on comparative investigation for various *o*-quinones. Throughout the investigation introduced in this book, first principles calculation based on density functional theory was employed for simulating the reactions. Moreover, as a model of the reaction environment in aqueous solutions, a continuous dielectric model that stabilizes polarized electronic states of molecules by means of dielectric response of solvent was used.

Chapter 2 introduces theoretical works performed on dopachrome conversion [123, 124], and Chap. 3 introduces theoretical studies regarding reactions of dopaquinone and *o*-quinones [125–127]. Chapter 4 summarizes the contents of this book and discusses future prospect.

References

1. M. d'Ischia, K. Wakamatsu, A. Napolitano, S. Briganti, J.C. García-Borrón, D. Kovacs, P. Meredith, A. Pezzella, M. Picardo, T. Sarna, J.D. Simon, S. Ito, Melanins and melanogenesis: methods standards protocols. Pigment Cell Melanoma Res. **26**, 616–633 (2013)
2. S. Ito, K. Wakamatsu, Chemistry of mixed melanogenesis-pivotal roles of dopaquinone. Photochem. Photobiol. **84**, 582–592 (2008)
3. S. Ito, K. Wakamatsu, Human hair melanins: what we have learned and have not learned from mouse coat color pigmentation. Pigment Cell Melanoma Res. **24**, 63–74 (2010)
4. M. Seiji, T.B. Fitzpatrick, R.T. Simpson, M.S.C. Birbeck, Chemical composition and terminology of specialized organelles (melanosomes and melanin granules) in mammalian melanocytes. Nature **197**, 1082–1084 (1963)
5. J.Y. Lin, D.E. Fisher, Melanocyte biology and skin pigmentation. Nature **445**, 843–850 (2007)
6. T.P. Dryja, M. O'Neil-Dryja, J.M. Pawelek, D.M. Albert, Demonstration of tyrosinase in the adult bovine uveal tract and retinal pigment epithelium. Invest. Ophthalmol. Visual Sci. **17**, 511–514 (1978)

7. L. Zecca, R. Pietra, C. Goj, C. Mecacci, D. Radice, E. Sabbioni, Iron and other metals in neuromelanins substantia nigra and putamen of human brain. J. Neurochem. **62**, 1097–1101 (1994)
8. R.J. D'Amato, Z.P. Lipman, S.H. Snyder, Selectivity of the Parkinsonian neurotoxin MPTP: toxic metabolite MPP + binds to neuromelanin. Science **231**, 987–989 (1986)
9. W. Westerhof, M. d'Ischia, Vitiligo puzzle: the pieces fall in place. Pigment Cell Res. **20**, 345–359 (2007)
10. P.B. Chapman et al., Improved survival with vemurafenib in melanoma with BRAF V600E mutation. N. Engl. J. Med. **364**, 2507–2516 (2011)
11. F.S. Hodi et al., Improved survival with ipilimumab in patients with metastatic melanoma. N. Engl. J. Med. **363**, 711–723 (2010)
12. R. Nazarian, H. Shi, Q. Wang, X. Kong, R.C. Koya, H. Lee, Z. Chen, M.-K. Lee, N. Attar, H. Sazegar, T. Chodon, S.F. Nelson, G. McArthur, J.A. Sosman, A. Ribas, R.S. Lo, Melanomas acquire resistance to B-RAF (V600E) inhibition by RTK or N-RAS upregulation. Nature **468**, 973–977 (2010)
13. R.M. MacKie, A. Hauschild, A.M.M. Eggermont, Epidemiology of invasive cutaneous melanoma. Annals Oncol. **20**, vi1-vi7 (2009)
14. P. Meredith, J. Riesz, Radiative relaxation quantum yields for synthetic eumelanin. Photochem. Photobiol. **79**, 211–216 (2004)
15. J.B. Nofsinger, T. Ye, J.D. Simon, Ultrafast nonradiative relaxation dynamics of eumelanin. J. Phys. Chem. B **105**, 2864–2866 (2001)
16. S. Subianto, G. Will, P. Meredith, Electrochemical synthesis of melanin free-standing films. Polymer **46**, 11505–11509 (2005)
17. J.E. McGinness, Mobility gaps: a mechanism for band gaps in Melanins. Science **177**, 896–897 (1972)
18. J.E. McGinness, P. Corry, P. Procter, Amorphous semiconductor switching in melanins. Science **183**, 853–855 (1974)
19. A.B. Mostert, B.J. Powell, I.R. Gentle, P. Meredith, On the origin of electrical conductivity in the bio-electronic material melanin. Appl. Phys. Lett. **100**, (2012)
20. A.B. Mostert, B.J. Powell, F.L. Pratt, G.R. Hanson, T. Sarna, I.R. Gentle, P. Meredith, Role of semiconductivity and ion transport in the electrical conduction of melanin. Proc. Natl. Acad. Sci. **109**, 8943–8947 (2012)
21. C.J. Bettinger, J.P. Bruggeman, A. Misra, J.T. Borenstein, R. Langer, Biocompatibility of biodegradable semiconducting melanin films for nerve tissue engineering. Biomaterials **30**, 3050–3057 (2009)
22. H. Lee, S.M. Dellatore, W.M. Miller, P.B. Messersmith, Mussel-inspired surface chemistry for multifunctional coatings. Science **318**, 426–430 (2007)
23. S. Ito, A Chemist's View of Melanogenesis. Pigment Cell Res. **16**, 230–236 (2003)
24. S. Ito, Y. Nakanishi, R.K. Valenzuela, M.H. Brilliant, L. Kolbe, K. Wakamatsu, Usefulness of alkaline hydrogen peroxide oxidation to analyze eumelanin and pheomelanin in various tissue samples: application to chemical analysis of human hair melanins. Pigment Cell Melanoma Res. **24**, 605–613 (2011)
25. H. Ozeki, K. Wakamatsu, S. Ito, Chemical characterization of eumelanins with special emphasis on 5,6-dihydroxyindole-2-carboxylic acid content and molecular size. Anal. Biochem. **248**, 149–157 (1997)
26. H. Ozeki, S. Ito, K. Wakamatsu, T. Hirobe, Chemical Characterization of hair melanins in various coat-color mutants of mice. J. Invest. Dermatol. **105**, 361–366 (1995)
27. V.J. Hearing, M. Jiménz, Mammalian tyrosinase: the critical regulatory control point in melanocyte pigmentation. Int. J. Biochem. **19**, 1141–1147 (1987)
28. T.B. Fitzpatrick, S.W. Becker Jr., A.B. Lerner, H. Montgomery, Tyrosinase in human skin: demonstration of its presence and of its role in human melanin formation. Science **112**, 223–225 (1950)
29. A.B. Lerner, T.B. Fitzpatrick, Biochemistry of melanin formation. Physiol. Rev. **30**, 91–126 (1950)

30. G.H. Hogeboom, M.H. Adams, Mammalian tyrosinase and dopa oxidase. J. Biol. Chem. **145**, 273–279 (1942)
31. A.B. Lerner, T.B. Fitzpatrick, E. Calkins, W.H. Summerson, Mammalian tyrosinase: preparation and properties. J. Biol. Chem. **178**, 185–195 (1949)
32. H.S. Raper, XCV. The tyrosinase-tyrosine reaction V. J. Biol. Chem. **20**, 735–742 (1926)
33. H.S. Raper, XIV. The tyrosinase-tyrosine reaction VI. J. Biol. Chem. **21**, 89–96 (1927)
34. K. Wakamatsu, S. Ito, Preparation of eumelanin-related metabolites 5,6-dihydroxyindole 5,6-dihydroxyindole-2-carboxylic acid and their *O*-methyl derivatives. Anal. Biochem. **170**, 335–340 (1988)
35. H.S. Mason, The chemistry of melanin. III. mechanism of the oxidation of dihydroxyphenylalanine by tyrosinase. J. Biol. Chem. **172**, 83–99 (1948)
36. J.D. Bu'Lock, J. Harley-Mason, Melanin and its precursors. Part II. Model experiments on the reactions between quinones and indoles and consideration of a possible structure for the melanin polymer. J. Chem. Soc. 703–712 (1951)
37. R.J.S. Beer, T. Broadhurst, A. Robertson, The chemistry of the melanins. Part V. The autoxidation of 5,6-dihydroxyindoles. J. Chem. Soc. 1947–1953 (1954)
38. T.B. Fitzpatrick, A.B. Lerner, Terminology of pigment cells. Science **117**, 640–645 (1953)
39. A.M. Körner, J.M. Pawelek, Dopachrome conversion: a possible control point in melanin biosynthesis. J. Invest. Dermatol. **75**, 192–195 (1980)
40. A.M. Körner, P. Gettins, Synthesis in vitro of 5,6-dihydroxyindole-2-carboxylic acid by dopachrome conversion factor from Cloudman S91 melanoma cells. J. Invest. Dermatol. **85**, 229–231 (1985)
41. S. Ito, Reexamination of the structure of eumelanin. Biochim. Biophys. Acta **883**, 155–161 (1986)
42. J.M. Pawelek, After dopachrome? Biochim. Biophys. Acta **883**, 155–161 (1986)
43. M. Sugumaran, V. Semensi, Quinone methide as a new intermediate in eumelanin biosynthesis. J. Biol. Chem. **266**, 6073–6078 (1991)
44. I.J. Jackson, D.M. Chambers, K. Tsukamoto, N.G. Copeland, D.J. Gilbert, N.A. Jenkins, V. Hearing, A second tyrosinase-related protein TRP-2 maps to and is mutated at the mouse slaty locus. EMBO J. **11**, 527–535 (1992)
45. F. Solano, J.H. Martinez-Liarte, C. Jiménz-Cervantes, J.C. García-Borrón, J.A. Lozano, Dopachrome tautomerase is a zinc-containing enzyme. Biochem. Biophys. Res. Commun. **204**, 1243–1250 (1994)
46. F. Solano, C. Jiménez-Cervantes, J.H. Martínez-Liarte, J.C. García-Borrón, J.R. Jara, J.A. Lozano, Molecular mechanism for catalysis by a new zinc-enzyme dopachrome tautomerase. Biochem. J. **313**, 447–453 (1996)
47. W.T. Ismaya, H.J. Rozeboom, A. Weijn, J.J. Mes, F. Fusetti, H.J. Wichers, B.W. Dijkstra, Crystal structure of *Agaricus bisporus* mushroom tyrosinase: identity of the tetramer subunits and interaction with tropolone. Biochemistry **50**, 5477–5486 (2011)
48. C.A. Ramsden, P.A. Riley, Tyrosinase: the four oxidation states of the active site and their relevance to enzymatic activation oxidation and inactivation. Bioorg. Med. Chem. **22**, 2388–2395 (2014)
49. S. Naish-Byfield, P.A. Riley, Oxidation of monohydric phenol substrates by tyrosinase. An oximetric study. Biochem. J. **288**, 63–67 (1992)
50. C.J. Cooksey, P.J. Garratt, E.J. Land, S. Pavel, C.A. Ramsden, P.A. Riley, N.P.M. Smit, Evidence of the indirect formation of the catecholic intermediate substrate responsible for the autoactivation kinetics of tyrosinase. J. Biol. Chem. **272**, 26226–26235 (1997)
51. E. Pelizzetti, E. Mentasti, E. Pramauro, G. Giraudi, Kinetic determination of adrenaline, L-dopa and their mixtures with a stopped-flow spectrophotometric technique. Anal. Chim. Acta **85**, 161–168 (1976)
52. D. Kertesz, M. Brunori, R. Zito, E. Antonini, Transient kinetic studies of dopa oxidation by polyphenoloxidase. Biochim. Biophys. Acta **250**, 306–310 (1971)
53. M.R. Chedekel, E.J. Land, A. Thompson, T.G. Truscott, Early steps in the free radical polymerisation of 3,4-dihydroxyphenylalanine (dopa) into melanin. J. Chem. Soc. Chem. Commun. 1170–1172 (1984)

54. R. Edge, M. d'Ischia, E.J. Land, A. Napolitano, S. Navaratnam, L. Panzella, A. Pezzella, C.A. Ramsden, P.A. Riley, Dopaquinone redox exchange with dihydroxyindole and dihydroxyindole carboxylic acid. Pigment Cell Res. **19**, 443–450 (2006)
55. C. Jímenez-Cervantes, F. Solano, T. Kobayashi, K. Urabe, V.J. Hearing, J.A. Lozano, J.C. García-Borrón, A new enzymatic function in the melanogenic pathway. The 5,6-dihydroxyindole-2-carboxylic acid oxidase activity of tyrosinase-related protein-1 (TRP1). J. Biol. Chem. **269**, 17993–18000 (1994)
56. T. Kobayashi, K. Urabe, A. Winder, C. Jímenez-Cervantes, G. Imokawa, T. Brewington, F. Solano, J.C. García-Borrón, V.J. Hearing, Tyrosinase related protein 1 (TRP1) functions as a DHICA oxidase in melanin biosynthesis. EMBO J. **13**, 5818–5825 (1994)
57. C. Olivares, C. Jímenez-Cervantes, J.A. Lozano, F. Solano, J.C. García-Borrón, The 5,6-dihydroxyindole-2-carboxylic acid (DHICA) oxidase activity of human tyrosinase. Biochem. J. **354**, 131–139 (2001)
58. S. Ito, N. Suzuki, S. Takebayashi, S. Commo, K. Wakamatsu, Neutral pH and copper ions promote eumelanogenesis after the dopachrome stage. Pigment Cell Melanoma Res. **26**, 817–825 (2013)
59. A. Napolitano, M.G. Corradini, G. Prota, A reinvestigation of the structure of melanochrome. Tetrahedron Lett. **26**, 2805–2808 (1985)
60. G. Prota, Melanins melanogenesis and melanocytes: looking at their functional significance from the chemist's viewpoint. Pigment Cell Res. **13**, 283–293 (2000)
61. A. Pezzella, A. Napolitano, M. d'Ischia, G. Prota, Oxidative polymerisation of 5,6-dihydroxyindole-2-carboxylic acid to melanin: a new insight. Tetrahedron **52**, 7913–7920 (1996)
62. P. Palumbo, M. d'Ischia, G. Prota, Tyrosinase-promoted oxidation of 5,6-dihydroxyindole-2-carboxylic acid to melanin. Isolation and characterization of oligomer intermediates, Tetrahedron **43**, 4203–4206 (1987)
63. P. Meredith, T. Sarna, The physical and chemical properties of eumelanin. Pigment Cell Res. **19**, 572–594 (2006)
64. M. d'Ischia, A. Napolitano, A. Pezzella, P. Meredith, T. Sarna, Chemical and structural diversity in eumelanins: unexplored bio-optoelectronic materials. Angew. Chem. Int. Ed. **48**, 3914–3921 (2009)
65. A. Pezzella, D. Vogna, G. Prota, Atropoisomeric melanin intermediates by oxidation of the melanogenic precursor 5,6-dihydroxyindole-2-carboxylic acid under biomimetic conditions. Tetrahedron **58**, 3681–3687 (2002)
66. L. Panzella, A. Pezzella, A. Napolitano, M. d'Ischia, The first 5,6-dihydroxyindole tetramer by oxidation of 5,5',6,6'-tetrahydroxy-2,4'-biindolyl and an unexpected issue of positional reactivity en route to eumelanin-related polymers. Org. Lett. **9**, 1411–1414 (2007)
67. A. Pezzella, L. Panzella, A. Natangelo, M. Arzillo, A. Napolitano, M. d'Ischia, 5,6-Dihydroxyindole tetramers with anomalous interunit bonding patterns by oxidative coupling of 5,5',6,6'-tetrahydroxy-2,7'-biindolyl: emerging complexities on the way toward an improved model of eumelanin buildup. J. Org. Chem. **72**, 9225–9230 (2007)
68. M. d'Ischia, A. Napolitano, A. Pezzella, E.J. Land, C.A. Ramsden, P.A. Riley, 5,6-Dihydroxyindoles and indole-5,6-diones. Adv. Heterocycl. Chem. **89**, 1–63 (2005)
69. A. Napolitano, O. Crescenzi, G. Prota, Copolymerisation of 5,6-dihydroxyindole and 5,6-dihydroxyindole-2-carboxylic Acid in melanogenesis: isolation of a cross-coupling produc. Tetrahedron Lett. **34**, 885–888 (1993)
70. A. Napolitano, O. Crescenzi, K. Tsiakas, G. Prota, Oxidation chemistry of 5,6-dihydroxy-2-methylindole. Tetrahedron **49**, 9143–9150 (1993)
71. K. Glass, S. Ito, P.R. Wilby, T. Sota, A. Nakamura, C.R. Bowers, J. Vinther, S. Dutta, R. Summons, D.E.G. Briggs, K. Wakamatsu, J.D. Simon, Direct chemical evidence for eumelanin pigment from the Jurassic period. Proc. Natl. Acad. Sci. **109**, 10218–10223 (2012)
72. S. Ito, K. Wakamatsu, K. Glass, J.D. Simon, High-performance liquid chromatography estimation of cross-linking of dihydroxyindole moiety in eumelanin. Anal. Biochem. **434**, 221–225 (2013)

73. H.C. Longuet-Higgins, On the origin of the free radical property of melanins. Arch. Biochem. Biophys. **86**, 231–232 (1960)
74. M.S. Blois, A.B. Zahlan, J.E. Maling, Electron spin resonance studies on melanin. Biophys. J. **4**, 471–490 (1964)
75. A. Pullman, B. Pullman, The band structure of melanins. Biochim. Biophys. Acta **54**, 384–385 (1961)
76. D.S. Galvão, M.J. Caldas, Polymerization of 5,6-indolequinone: a view into the band structure of melanins. J. Chem. Phys. **88**, 4088–4091 (1988)
77. D.S. Galvão, M.J. Caldas, Theoretical investigation of model polymers for eumelanins. I. Finite and infinite polymers. J. Chem. Phys. **92**, 2630–2636 (1990)
78. J. Cheng, S.C. Moss, M. Eisner, P. Zschack, X-ray characterization of melanins–I. Pigment Cell Res. **7**, 255–262 (1994)
79. J. Cheng, S.C. Moss, M. Eisner, X-ray characterization of melanins–II. Pigment Cell Res. **7**, 263–273 (1994)
80. G.W. Zajac, J.M. Gallas, J. Cheng, M. Eisner, S.C. Moss, A.E. Alvarado-Swaisgood, The fundamental unit of synthetic melanin: a verification by tunneling microscopy of X-ray scattering results. Biochim. Biophys. Acta **1199**, 271–278 (1994)
81. P. Meredith, B.J. Powell, J. Riesz, S.P. Nighswander-Rempel, M.R. Pederson, E.G. Moore, Towards structure-property-function relationships for eumelanin. Soft Matter **2**, 37–44 (2006)
82. P. Hohenberg, W. Kohn, Inhomogeneous electron gas. Phys. Rev. **136**, B864–B871 (1964)
83. W. Kohn, L.J. Sham, Self-consistent equations including exchange and correlation effects. Phys. Rev. **140**, A1133–A1138 (1965)
84. Y.V. Il'ichev, J.D. Simon, Building blocks of eumelanin: relative stability and excitation energies of tautomers of 5,6-dihydroxyindole and 5,6-indolequinone. J. Phys. Chem. B **107**, 7162–7171 (2003)
85. B.J. Powell, T. Baruah, N. Bernstein, K. Brake, R.H. McKenzie, P. Meredith, M.R. Pederson, A first-principles density-functional calculation of the electronic and vibrational structure of the key melanin monomers. J. Chem. Phys. **120**, 8608–8615 (2004)
86. B.J. Powell, 5,6-Dihydroxyindole-2-carboxylic acid: a first principles density functional study. Chem. Phys. Lett. **402**, 111–115 (2005)
87. M.L. Tran, B.J. Powell, P. Meredith, Chemical and structural disorder in eumelanins: a possible explanation for broadband absorbance. Biophys. J. **90**, 743–752 (2006)
88. J.S.M. Anderson, J. Melin, P.W. Ayers, Conceptual density-functional theory for general chemical reactions including those that are neither charge-nor frontier-orbital controlled. 1. Theory and derivation of a general-purpose reactivity indicator. J. Chem. Theory Comput. **3**, 358–374 (2007)
89. J.S.M. Anderson, J. Melin, P.W. Ayers, Conceptual density-functional theory for general chemical reactions including those that are neither charge-nor frontier-orbital controlled. 2. Application to molecules where frontier molecular orbital theory fails. J. Chem. Theory Comput. **3**, 375–389 (2007)
90. H. Okuda, K. Wakamatsu, S. Ito, T. Sota, Possible oxidative polymerization mechanism of 5,6-dihydroxyindole from ab initio calculations. J. Phys. Chem. A **112**, 11213–11222 (2008)
91. R.G. Parr, W. Yang, Density functional approatch to the frontier-electron theory of chemical reactivity. J. Am. Chem. Soc. **106**, 4049–4050 (1984)
92. P. Fuentealba, P. Pérez, R. Contreras, On the condensed Fukui function. J. Chem. Phys. **113**, 2544–2551 (2000)
93. S. Ito, G. Prota, A facile one-step synthesis of cysteinyldopas using mushroom tyrosinase. Experimentia **33**, 1118–1119 (1977)
94. A. Thompson, E.J. Land, M.R. Chedekel, K.V. Subbarao, T.G. Truscott, A pulse radiolysis investigation of the oxidation of the melanin precursors 3,4-dihydroxyphenylalanine (dopa) and the cysteinyldopas. Biochim. Biophys. Acta **843**, 49–57 (1985)
95. A. Napolitano, P.D. Donato, G. Prota, E.J. Land, Transient quinonimines and 1,4-benzothiazines of pheomelanogensis: new pulse radiolytic and spectrophotometric evidence. Free Radic. Biol. Med. **27**, 521–528 (1999)

96. A. Napolitano, P.D. Donato, G. Prota, New regulatory mechanisms in the biosynthesis of pheomelanins: rearrangement versus redox exchange reaction routes of a transient 2H-1,4-benzothiazine-*o*-quinonimine intermediate. Biochim. Biophys. Acta **1475**, 47–54 (2000)
97. A. Napolitano, C. Costantini, O. Crescenzi, G. Prota, Characterisation of 1,4-benzothiazine intermediates in the oxidative conversion of 5-*S*-cysteinyldopa to pheomelanin. Tetrahedron Lett. **35**, 6365–6368 (1994)
98. A. Napolitano, M.D. Lucia, L. Panzella, M. d'Ischia, The "benzothiazine" chromophore of pheomelanins: a reassessment. Photochem. Photobiol. **84**, 593–599 (2008)
99. K. Wakamatsu, K. Ohtara, S. Ito, Chemical analysis of late stages of pheomelanogenesis: conversion of dihydrobenzothiazine to a benzothiazole structure. Pigment Cell Melanoma Res. **22**, 474–486 (2009)
100. E.J. Land, C.A. Ramsden, P.A. Riley, Pulse radiolysis studies of *ortho*-quinone chemistry relevant to melanogenesis. J. Photochem. Photobiol. B Biol. **64**, 123–135 (2001)
101. A. Napolitano, P.D. Donato, G. Prota, Zinc-catalyzed oxidation of 5-S-cysteinyldopa to 2,2'-bi(2H-1,4-benzothiazine): tracing the biosynthetic pathway of trichochromes the characteristic pigments of red hair. J. Org. Chem. **66**, 6958–6966 (2001)
102. P.D. Donato, A. Napolitano, G. Prota, Metal ions as potential regulatory factors in the biosynthesis of red hair pigments: a new benzothiazole intermediate in the iron or copper assisted oxidation of 5-*S*-cysteinyldopa. Biochim. Biophys. Acta **1571**, 157–166 (2002)
103. A. Biesemeier, U. Schraermeyer, O. Eibl, Chemical composition of melanosomes lipofuscin and melanolipofuscin granules of human RPE tissues. Exp. Eye Res. **93**, 29–39 (2011)
104. Y. Liu, L. Hong, K. Wakamatsu, S. Ito, B. Adhyaru, C.Y. Cheng, C.R. Bowers, J.D. Simon, Comparison of structural and chemical properties of black and red human hair melanosomes. Photochem. Photobiol. **81**, 135–144 (2005)
105. K. Thörneby-Andersson, O. Sterner, C. Hansson, Tyrosinase-mediated formation of a reactive quinone from the depigmenting agents 4-*tert*-butylphenol and 4-*tert*-butylcatechol. Pigment Cell Res. **13**, 33–38 (2000)
106. P. Manini, A. Napolitano, W. Westerhof, P.A. Riley, M. d'Ischia, A reactive *ortho*-quinone generated by tyrosinase-catalyzed oxidation of the skin depigmentating agent monobenzone: self-coupling and thiol-conjugation reactions and possible implications for melanocyte toxicity. Chem. Res. Toxicol. **22**, 1398–1405 (2009)
107. K. Hasegawa, S. Ito, S. Inoue, K. Wakamatsu, H. Ozeki, I. Ishiguro, Dihydro-1,4-benzothiazine-6,7-dione, the ultimate toxic metabolite of 4-*S*-cysteaminylphenol and 4-*S*-cysteaminylcatechol. Biochem. Pharmacol. **53**, 1435–1444 (1997)
108. S. Ito, M. Ojika, T. Yamashita, K. Wakamatsu, Tyrosinase-catalyzed oxidation of rhododendrol produces 2-methylchromane-6,7-dione, the putative ultimate toxic metabolite: implications for melanocyte toxicity. Pigment Cell Melanoma Res. **27**, 744–753 (2014)
109. S. Ito, W. Gerwat, L. Kolbe, T. Yamashita, M. Ojika, K. Wakamatsu, Human tyrosinase is able to oxidize both enantiomers of rhododendrol. Pigment Cell Melanoma Res. **27**, 1149–1153 (2014)
110. V. Hariharan, J. Klarquist, M.J. Reust, A. Koshoffer, M.D. McKee, R.E. Boissy, I.C. Le Poole, Monobenzyl ether of hydroquinone and 4-tertiary butyl phenol activate markedly different physiological responses in melanocytes: relevance to skin depigmentation. J. Invest. Dermatol. **130**, 211–220 (2010)
111. S. Toosi, S.J. Orlow, P. Manga, Vitiligo-inducing phenols activate the unfolded protein response in melanocytes resulting in upregulation of IL6 and IL8. J. Invest. Dermatol. **132**, 2601–2609 (2012)
112. F. Yang, R. Sarangarajan, I.C. Le Poole, E.E. Medrano, R.E. Boissy, The cytotoxicity and apoptosis induced by 4-tertiary butylphenol in human melanocytes are independent of tyrosinase activity. J. Invest. Dermatol. **114**, 157–164 (2000)
113. J.G. van den Boorn, D.I. Picavet, P.F. van Swieten, H.A. van Veen, D. Konijnenberg, P.A. van Veelen, T. van Capel, E.C. de Jong, E.A. Reits, J.W. Drijfhout, J.D. Bos, C.J.M. Melief, R.M. Luiten, Skin-depigmenting agent monobenzone induces potent T-cell autoimmunity toward pigmented cells by tyrosinase haptenation and melanosome autophagy. J. Invest. Dermatol. **131**, 1240–1251 (2011)

114. Y. Ishii-Osai, T. Yamashita, Y. Tamura, N. Sato, A. Ito, H. Honda, K. Wakamatsu, S. Ito, E. Nakayama, M. Okura, K. Jimbow, *N*-propionyl-4-S-cysteaminylphenol induces apoptosis in B16F1 cells and mediates tumor-specific T-cell immune responses in a mouse melanoma model. J. Dermatol. Sci. **67**, 51–60 (2012)
115. M. Sasaki, M. Kondo, K. Sato, M. Umeda, K. Kawabata, Y. Takahashi, T. Suzuki, K. Matunaga, S. Inoue, Rhododendrol a depigmentation-inducing phenolic compound exerts melanocyte cytotoxicity via a tyrosinase-dependent mechanism. Pigment Cell Melanoma Res. **27**, 754–763 (2014)
116. E. Karg, E. Rosengren, H. Rorsman, Hydrogen peroxide as a mediator of dopac-induced effects on melanoma cells. J. Invest. Dermatol. **96**, 224–227 (1991)
117. S. Ito, M. Okura, Y. Nakanishi, M. Ojika, K. Wakamatsu, T. Yamashita, Tyrosinase-catalyzed metabolism of rhododendrol (RD) in B16 melanoma cells: production of RD-pheomelanin and covalent binding with thiol proteins. Pigment Cell Melanoma Res. **28**, 295–306 (2015)
118. A. Napolitano, L. Panzella, G. Monfrecola, Pheomelanin-induced oxidative stress: bright and dark chemistry bridging red hair phenotype and melanoma. Pigment Cell Melanoma Res. **27**, 721–733 (2014)
119. L. Panzella, L. Leone, G. Greco, G. Vitiello, G. D'Errico, A. Napolitano, M. d'Ischia, Red human hair pheomelanin is a potent pro-oxidant mediating UV-independent contributory mechanisms of melanomagenesis. Photochem. Photobiol. **82**, 733–737 (2006)
120. T. Ye, L. Hong, J. Garguilo, A. Pawlak, G.S. Edwards, R.J. Nemanich, T. Sarna, J.D. Simon, Photoionization thresholds of melanins obtained from free electron laser-photoelectron emission microscopy femtosecond transient absorption spectroscopy and electron paramagnetic resonance measurements of oxygen photoconsumption. Pigment Cell Melanoma Res. **28**, 295–306 (2015)
121. W. Westerhof, P. Manini, A. Napolitano, M. d'Ischia, The haptenation theory of vitiligo and melanoma rejection: a close-up. Exp. Dermatol. **20**, 92–96 (2011)
122. T. Passeron, J.-P. Ortonne, Activation of the unfolded protein response in vitiligo: the missing link? J. Invest. Dermatol. **132**, 2502–2504 (2012)
123. R. Kishida, Y. Ushijima, A.G. Saputro, H. Kasai, Effect of pH on elementary steps of dopachrome conversion from first-principles calculation. Pigment Cell Melanoma Res. **27**, 734–743 (2014)
124. R. Kishida, A.G. Saputro, H. Kasai, Mechanism of dopachrome tautomerization into 5,6-dihydroxyindole-2-carboxylic acid catalyzed by Cu(II) based on quantum chemical calculations. Biochim. Biophys. Acta **2015**, 281–286 (1850)
125. R. Kishida, H. Kasai, S.M. Aspera, R.L. Arevalo, H. Nakanishi, Branching reaction in melanogenesis: the effect of intramolecular cyclization on thiol binding. J. Electron. Mater. **46**, 3784–3788 (2017)
126. R. Kishida, H. Kasai, S.M. Aspera, R.L. Arevalo, H. Nakanishi, Density functional theory-based first principles calculations of rhododendrol-quinone reactions: preference to thiol binding over cyclization. J. Phys. Soc. Jpn. **86**, 024804-1-5 (2017)
127. R. Kishida, A.G. Saputro, R.L. Arevalo, H. Kasai, Effects of introduction of α-carboxylate *N*-methyl and *N*-formyl groups on intramolecular cyclization of *o*-quinone amines: density functional theory-based study. Int. J. Quant. Chem. **117**, e25445-1-9 (2017)

Chapter 2
Dopachrome Conversion

Abstract The color and redox properties of eumelanins are affected by the presence of the carboxyl groups in their indollic structural unit. The key reaction determining the amount of carboxyl groups in eumelanin is dopachrome conversion. In this conversion, 5,6-dihydroxylindole (DHI) is spontaneously formed through the elimination of the carboxyl group, namely decarboxylation, whereas the non-decarboxylated product 5,6-dihydroxylindole-2-carboxylic acid (DHICA) is also formed in the presence of several factors such as dopachrome tautomerase (DCT) and Cu(II) ion as well as strongly alkaline pH. In this chapter, we introduce computational studies of dopachrome conversion with the emphasis on the branching into DHI and DHICA formation. As a result, important factors affecting the selective formation of DHI and DHICA were identified. These reaction modes are switched based on the protonation/deprotonation of the quinonoid group of dopachrome. The catalytic effects of basic pH and Cu(II), experimentally observed, can be explained based on two aspects: (i) promotion of the rate-determining step, namely β-deprotonation and (ii) protection of quinonoid group from protonation. Our approach clarifies dopachrome conversion from atomic nuclei and electrons world.

Keywords Dopachrome · 5,6-dihydroxyindole (DHI) · 5,6-dihydroxyindole-2-carboxylic acid (DHICA) · Deprotonation · Cu(II) coordination · Density functional theory

2.1 Introduction

2.1.1 Background on Dopachrome Studies

Dopachrome conversion produces DHI and DHICA, which are the monomers of eumelanin, and thus directly determines the properties of the eumelanin produced (Fig. 2.1). It has been pointed out that eumelanin may not only protect the skin and the hair from UV radiation but also play an important role in the scavenging of intracellular ROS [1–3]. This antioxidant effect is likely to be derived from the DHICA unit contained in eumelanin. Furthermore, DHICA itself or its methoxy derivatives may also have physiologically important effects, including antioxidant

R. Kishida et al., *Melanin Chemistry Explored by Quantum Mechanics*,
https://doi.org/10.1007/978-981-16-1315-9_2

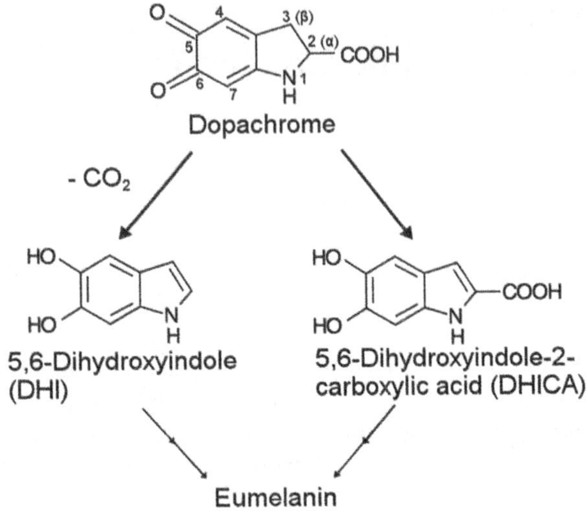

Fig. 2.1 Conversion of dopachrome into DHI or DHICA. Numbering is based on IUPAC nomenclature

activity [4, 5]. These suggest a physiological significance of controlling dopachrome conversion to produce DHICA.

So far, various factors that could control dopachrome conversion have been reported. At room temperature, atmospheric pressure, and physiological pH (around 7.0), without enzymes, more than 95% of dopachrome slowly and spontaneously converts to DHI (with rate constant 4.0×10^{-4} s^{-1}). The preference of dopachrome conversion mainly depends on pH, metal ions, and the enzyme DCT.

Muneta used a tyrosinase-like enzyme isolated from potatoes to investigate the pH dependence of the conversion rate of dopachrome using spectrophotometry [6]. Results show that dopachrome converted more rapidly at neutral pH (7.0) than weakly acidic pH (5.0). This result was reproduced by a chemical experiment performed by Ito et al. [7], where the rate constant at pH 7.3 was 4.6 times higher than that at pH 5.3. In addition, chemical degradation analyses showed that the DHI/DHICA ratio was not affected at this pH range. DHICA production is preferred under extremely acidic or basic conditions, which are not physiological conditions. Mason, who identified dopachrome, reported selective DHICA production at pH 1.3–2.0 [8]. Stravs-Mombelli and Wyler have showed that DHICA production is preferred at pH 13. Wakamatsu and Ito reproduced this reaction at the strongly basic pH and showed that this condition gave 12 of DHICA/DHI ratio [9].

Palumbo et al. investigated the effects of metal ions as a catalyst for dopachrome conversion [10]. The metal ions used were Fe(III), Al(III), Ca(II), Mn(II), Zn(II), Co(II), Ni(II), and Cu(II). These metal ions promoted dopachrome conversion, and the rate constant increased linearly in a first-order manner with respect to the metal ion concentration. Among them, Cu(II), Ni(II), and Co(II) have especially high catalytic activity (listed in the order of increasing activity). Moreover, these three metal ions selectively catalyzed the formation of DHICA.

The catalytic activity of Zn(II) on dopachrome conversion was relatively very weak. This is a little surprising if one should consider the fact that DCT contained Zn(II) at the active sites. A reexamination given by Ito et al. reproduced the catalytic activity of Cu(II) on dopachrome conversion, and also showed that Cu(II) promoted the oxidative polymerization to form eumelanin with higher ratios of DHICA [7]. Since Cu(II) is relatively abundant in melanosomes [11, 12], the DHI/DHICA ratio of eumelanin may be significantly affected by not only DCT but also Cu(II). Palumbo conducted a comparative experiment for the two catalytic factors, DCT and Cu(II), on dopachrome conversion [13]. Their results show that DCT had higher catalytic activity than Cu(II).

From the above experimental results, the main factors that control dopachrome conversion are pH (of melanosomes) and the DCT activity, while Cu(II) may also act as secondary factor. It was also reported that correlation between the DCT activity and the DHI/DHICA ratio could be not straightforward [14, 15]. Commo et al. showed that human follicular melanocytes from elderly individuals (aged older than 45 years old) have scarcely detectable DCT proteins. Nevertheless, melanin from these samples showed relatively high DHICA content (33–45%), clearly demonstrating the existence of an alternative mechanism to convert dopachrome to DHICA that does not rely on DCT [15, 16]. In other words, Cu(II) ions in melanosomes are likely to be a complementary factor, which promote DHICA production even in the absence of DCT activity.

In the case of DCT-catalyzed reaction, the selective DHICA production may be explained by a relatively straightforward mechanism. From the reported high stereospecificity, DCT presumably has a site capable of recognizing carboxyl group of dopachrome [17]. (Note that this carboxyl group is located at a chiral carbon.) Therefore, the formation of DHI would be suppressed by the inhibition of decarboxylation due to the interaction with the carboxyl-recognizing site of DCT. In contrast, it is not clear how (non-protein bound) metal ions such as Cu(II) alone catalyzes selective formation of DHICA. Although the metal ion-catalyzed DHICA formation was thought to occur through chelation of metal ions at the quinonoid site of dopachrome, mechanistic significance of such metal-dopachrome complexes on the DHICA formation is unclear.

pH is another factor affecting dopachrome conversion, although its mechanistic roles have not yet been clarified. The reported slower conversion at acidic pH would be due to the suppression of the rate-limiting deprotonation of dopachrome. Vavricka et al. showed that the conversion rate was also correlated with the concentration of buffer solution even at the same pH [18]. Interestingly, some types of buffer solution promoted DHICA production. From these results, there would be various factors besides pH affecting proton exchange processes between dopachrome and solvent molecules. Furthermore, the reported preference of DHICA formation under strongly acidic and basic conditions are also not clear.

Dopachrome conversion is a reaction that proceeds through proton rearrangements and forms a transient unstable species. There have been experimental difficulties in investigating the behavior of protons at the level of elementary reactions. To elucidate the conversion mechanism, Sugumaran et al. prepared esterified dopachrome to protect the carboxyl group, and then tried to identify possible intermediates during the conversion [19, 20]. Results show that an intermediate with a quinone methide structure was identified by HPLC analysis, both in the cases of non-enzymatic and enzymatic conditions.

2.1.2 *Computational Study on Dopachrome Conversion*

As reviewed above, progress on the understanding of the chemical nature of dopachrome conversion has been advancing from the early phase of melanin chemistry study. Especially, the importance of the proton rearrangement processes on dopachrome conversion was indicated by several studies as influenced by pH and type of buffer solution. Furthermore, coordinate bond formation between metal ions and the quinonoid group of dopachrome may play a crucial role in the selective formation of DHICA.

To understand the conversion mechanism of dopachrome and the catalytic effects of metal ions, we conducted a density functional theory-based first principles calculation on various molecular structures of prototropic isomers involved in dopachrome conversion [21, 22]. Here, we investigated Cu(II) as the catalytic factor of dopachrome conversion. Three elementary processes in dopachrome conversion were identified (Fig. 2.2). One is α-deprotonation, a process necessary for DHICA formation. Another is β-deprotonation, which corresponds to the formation of the quinone methide intermediate. The elementary process required for DHI formation is decarboxylation. Thus, we calculated and compared the activation barrier for these reactions in order to discuss how dopachrome conversion proceeds.

The following discussions were arranged in each section: Sect. 2.2 describes the calculation method and the model structure to be used. Sects. 2.3 and 2.4 describe the computational results of dopachrome conversion without and with Cu(II) coordination, respectively. And, Sect. 2.5 depicts the proposed scheme of dopachrome conversion.

Briefly, the following findings were obtained from the calculation results. The reaction starts mainly from β-deprotonation at nearly neutral pH in the absence

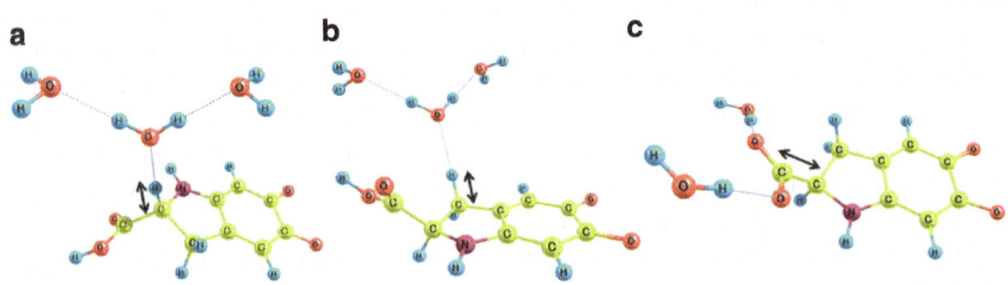

Fig. 2.2 Initial structures for calculation of the activation barriers for **a** α-deprotonation, **b** β-deprotonation, and **c** decarboxylation

and presence of Cu(II) coordination at the quinonoid group (5,6-carbonyl groups) of dopachrome. When Cu(II) is coordinated to the quinonoid group, the activation barriers for α-deprotonation, β-deprotonation, and decarboxylation are all reduced. Without Cu(II) coordination, the dissociated proton (from β-carbon) would be reprotonated at 5-carbon to form the metastable quinone methide intermediate. This quinone methide intermediate shows significantly reduced activation barriers for decarboxylation and α-deprotonation, as compared to those of the initial state of the reaction. When the 6-oxygen of the quinone methide is further protonated, the activation barrier for decarboxylation is drastically reduced, and then DHI is formed as the product. At a basic pH, this O6-protonation rate must decrease, and α-deprotonation rate should increase instead, generally indicating that this is a favorable condition for the formation of DHICA. In the presence of Cu(II) coordination at the quinonoid group, this reprotonation becomes energetically favorable at α-carboxyl group rather than at quinonoid group. This results in selective formation of DHICA. Since the rate-limiting step of dopachrome conversion is β-deprotonation based on its activation barriers with and without Cu(II) coordination, our calculated results confirm the reported base-catalyzed nature of this reaction. These calculations emphasize that protection of the quinonoid group from protonation is important for the selective formation of DHICA.

2.2 Calculation Methods and Models for Simulating Dopachrome Conversion

We conducted first principles calculations based on density functional theory [23, 24]. All calculations were performed using Gaussian09, which is a widely used quantum chemical calculation package [25]. The exchange correlation energy was calculated using a hybrid functional B3LYP [26, 27] and the basis set was 6–31 ++G(d, p). The natural atomic orbital analysis was performed to estimate the atomic charge [28].

Furthermore, the interaction with water was described using a polarizable continuum model (PCM) [29, 30]. To calculate the solvation energy by means of dielectric response, PCM approximates the solvent as a continuous dielectric medium with spherical cavities, which are introduced around the solute atoms (The cavity volume is approximately 1.1 times larger than the van der Waals volume). At the boundary between the cavities (vacuum) and the dielectric medium (water), the dielectric constant changes discontinuously so that apparent surface charges appear. The solvation energy can be calculated based on the interaction between this surface charge and the calculated electron density.

In order to calculate the activation barriers for elementary processes, potential energy curves are calculated along the direction in which the C–H or C–C bond length increases with a step size increment of 0.05 or 0.10 Å. For each point of the potential energy curve, the molecular geometry was optimized except for the dissociating bond length. However, in the cases of decarboxylation, "immediate rotation" of the dissociating carboxyl group was found several times. "Immediate rotation" here means an artifactual phenomenon in which the dihedral angle between dopachrome and the dissociating CO_2 (O=C–C2–C1) suddenly changes by approximately 90° with 0.05 Å of the bond length increment through the geometrical optimization. To avoid artifactual underestimation of the activation barrier, the corresponding dihedral angle was set to be frozen from the point of immediate rotation.

Deprotonation proceeds with the formation of hydronium ion H_3O^+. In this process, water molecule(s) act not only as a dielectric medium but also as a direct acceptor for the dissociating proton. Therefore, in addition to PCM, we directly put three H_2O molecules around the dissociating proton for the calculation of the activation barriers (Fig. 2.2). H_3O^+ can form hydrogen bonds with three H_2O molecules at the maximum. Immediately after the deprotonation, the generated H_3O^+ is likely to form two hydrogen bonds with surrounding H_2O molecules because the hydrogen that came from dopachrome is not directly facing the H_2O molecules.

To complete decarboxylation, the negatively charged carboxylate ion must become electrically neutral. This change in charge state of the carboxyl group gives significantly different hydrogen bond strength. In order to incorporate this effect in the calculation, we also put H_2O molecules around the dissociating carboxyl group. Specifically, two H_2O molecules are placed near the carboxyl oxygens (Fig. 2.2).

Possible sites of Cu(II) coordination are the quinonoid group (5,6-oxygen) and the (α-) carboxyl group (Fig. 2.3). Our preliminary calculation found only slight difference in total energy; the quinonoid coordination was slightly more stable with the energy difference of −0.86 kcal/mol. Thus, carboxyl coordination of Cu(II) cannot be ruled out in principle. Nevertheless, unlike the case of quinonoid coordination, this carboxyl group is σ-bonded with α-carbon. Therefore, Cu(II) coordination at this site cannot strongly electronically influence the π-conjugated system, which includes α-carbon and β-carbon, so that deprotonation from these sites would not be also affected by the presence of the Cu(II). Our preliminary calculation confirmed this hypothesis; the calculated activation barrier for β-deprotonation from the carboxyl coordinated Cu(II)-dopachrome was comparable with that of without Cu(II) coordination (Data not shown). From this point, the major catalytic effect of Cu(II) is likely to come from the quinonoid coordination, although the carboxyl coordination may

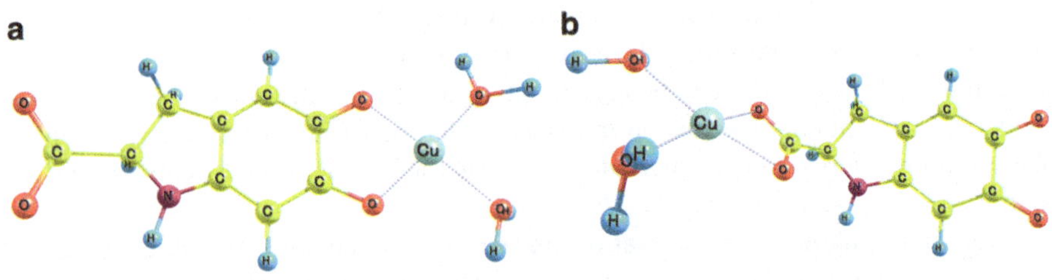

Fig. 2.3 Cu(II) coordination to **a** quinonoid group and **b** carboxyl group. Reprinted (with minor modification) from Ref. [22] with permission from Elsevier

also complementarily participate in the selective conversion to DHICA by inhibiting decarboxylation as a minor factor.

For the possible coordination geometry of Cu(II)-dopachrome complex, a four-coordinate model was used. In this structure, Cu(II) is planarly bound with the quinonoid group and two H_2O molecules, as shown in Fig. 2.3. Although Cu(II) aqua complex may be present in a five-fold-coordinated structure [31, 32], one of the coordination with H_2O is very weak (weaker than the hydrogen bond formed in the H_2O molecular cluster). Since this weak coordination shows fluctuating behavior of the coordination position, we did not include this in our model.

To calculate the Gibbs free energy, we first computed the Hessian matrix, and then diagonalized it to find the normal mode frequencies of the molecule. A partition function was obtained by considering the degrees of freedom for the molecular vibration, rotation, and translation. The temperature was set to 309.5 K considering the human body temperature, and the pressure was set to 1.0 atm. This thermodynamic model assumes a non-interacting dopachrome ideal gas (with PCM correction), and also that the pressure–volume product is uniquely determined by the temperature.

Theoretical methods for calculating the solvation free energy are still in the developing stages and the complete description cannot be expected in the short term. Although we used PCM for the qualitative description of the solvent–solute interaction, this can be extended into the following model; the solute molecules (dopachrome or dopachrome-Cu(II) complex) are surrounded by H_2O molecules, and form several hydration layers. This hydration layer is considered to have a strong interaction with the solute molecules so that the position of the O atoms in H_2O do not significantly change by thermal fluctuation.

Keeping in mind the presence of the above-mentioned hydration layers, we also conducted calculations with several H_2O molecules when considering deprotonations, decarboxylation, and complex formation between Cu(II) and dopachrome. The entire solution is divided into two liquid phase regions: a liquid phase region including at least solute molecules and the corresponding hydration layer, and a liquid phase region surrounding the hydrated region. (As described above, the solute concentration is sufficiently low so that the distances between solute molecules are large enough to ignore the interactions.) It is assumed that the heat exchange, expansion and compression work, and proton exchange are possible between the two liquid phase regions, and have the same temperature, pressure, and pH.

In this model, the protonations and deprotonations are affected by pH. Since pH determines the chemical potential of protons, the deprotonated states become stable at high pH, and the protonated states are favorable at low pH. The major protonated/deprotonated states in the equilibrium state under a given pH are determined by the acid dissociation constant K_a of the corresponding functional group. The concentrations of the protonated and deprotonated states become almost identical at the condition $pH = pK_a$ as described by Henderson-Hasselbalch equation. For example, to calculate the acid dissociation constant of the carboxyl group in dopachrome, it is necessary to compute the Gibbs free energy change for the proton dissociation. For this purpose, we considered a thermodynamic cycle as shown in Fig. 2.4.

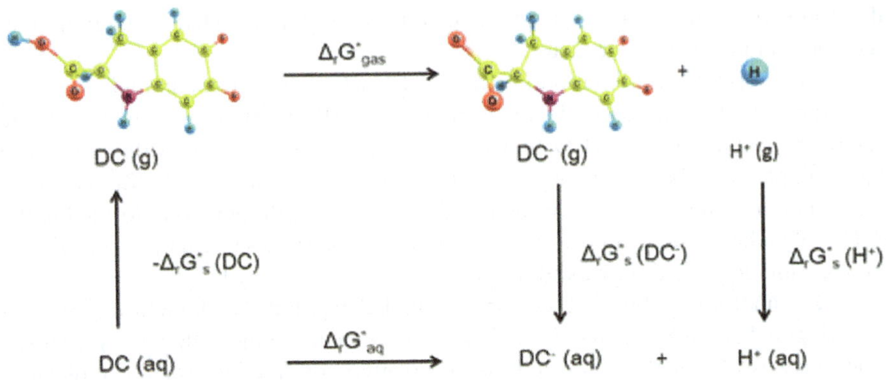

Fig. 2.4 Thermochemical cycle for carboxy deprotonation from dopachrome (DC). The Gibbs free energy of the carboxy deprotonation in aqueous solution $\Delta_r G^*_{aq}$ was calculated based on this cycle. States in gas phase and aqueous solution are denoted as g and aq in parentheses, respectively. $\Delta_r G^*_{gas}$ and $\Delta_r G^*_s$ denote the Gibbs free energy of the carboxy deprotonation in gas phase and of the hydration, respectively. Reprinted (with minor modification) from Ref. [21] with permission from Wiley

PCM alone does not guarantee sufficient precision for the quantitative calculation of the pK_a of dopachrome. Gaussian09 recommends the use of a semi-empirical solvation model called SMD for the quantitative calculation of the solvation free energy [33]. Thus, we used SMD only for the solvation free energy calculations. Nevertheless, even using SMD, it is still difficult to calculate the free energy for proton hydration, partly because of the very strong proton–water interaction and the small mass of proton that makes the quantum effects more evident. Therefore, we exceptionally use an experimental value of the free energy for proton hydration 265.75 kcal/mol (at 309.5 K) [34]. By using SMD and the experimental value, we obtained 1.99 of pK_a. This value is close to the carboxyl group of amino acids [35]. From this value, the carboxyl group should be present in the proton-dissociated state at physiological pH.

2.3 Dopachrome Conversion Mechanism Without Cu(II) Coordination

Dopachrome has four proton accepting groups, namely carboxyl group, amino group, and two quinonoid carbonyl groups (at 5-oxygen and 6-oxygen). At the electrically neutral condition, two of these four groups are protonated. Here, we compared the energetic stability for five prototropic isomers (A–E) as listed in Table 2.1.

The calculated results without PCM show that the carboxyl- and O6-protonated structure [C (vac.) defined in Table 2.1] is the energetically most stable. On the other hand, when calculated using PCM, the carboxyl- and amino-protonated structure

Table 2.1 Energetic stability of dopachrome prototropic tautomers at the initial step

Tautomer[a]	Protonation sites[b]	Energy (kcal/mol)[c]	Gibbs free energy (kcal/mol)[d]	Equilibrium composition[e]
A (vac.)	Carboxyl, N1	0.0	0.0	4.4×10^{-4}
B (vac.)	N1, O6	11.3	11.7	2.3×10^{-12}
C (vac.)	Carboxyl, O6	−5.5	−4.8	1.0
D (vac.)	N1, O5	28.0	27.5	1.7×10^{-23}
E (vac.)	Carboxyl, O5	Unstable[f]	Unstable[f]	0.0
A (aq.)	Carboxyl, N1	−18.5	−18.6	1.0
B (aq.)	N1, O6	−11.8	−11.4	8.4×10^{-6}
C (aq.)	Carboxyl, O6	−15.4	−15.4	6.1×10^{-3}
D (aq.)	N1, O5	0.7	0.7	2.6×10^{-14}
E (aq.)	Carboxyl, O5	0.3	−0.5	1.8×10^{-13}

[a]Symbols of dopachrome prototropic tautomers. Calculation without and with PCM is, respectively, denoted as (vac.) and (aq.)
[b]Numbers in this column correspond to the labels in Fig. 2.1
[c]The energy origin was set to the value of A (vac.)
[d]The energy origin was set to the value of A (vac.). Temperature was set to 309.5 K as a condition of human body
[e]Equilibrium composition is defined as the mole fraction of each tautomer in equilibrium state, normalized by amount of all tautomers in vacuo or in aqueous solution. These compositions were calculated based on the values of the Gibbs free energies. 1.0 of the activity coefficient was used as an approximate value
[f]Spontaneous proton transfer to O6 occurred

[A (aq.) defined in Table 2.1] was found to be the most stable. The electric dipole moment of this isomer A (aq.) was 14.7 D, while the isomer C (aq.) shows a smaller dipole moment 5.7 D. From this point, the isomer A can be said to have an electronic structure that is greatly influenced by dopachrome–water dielectric interaction. Thus, we consider that the electroneutral dopachrome prefers the carboxyl- and amino-protonated structure A as the initial state of conversion. Although the structure A is protonated at the carboxyl group, the estimated pK_a of this carboxyl group is 2.0 (see Sect. 2.2) so that this group must be deprotonated at physiological pH. In other words, it can be said that the energetic preference of the structure A does not contribute to the inhibition of decarboxylation.

Based on the identified initial structure (i.e. the structure A), we calculated the activation barriers for α-deprotonation, β-deprotonation, and decarboxylation to determine the initial step of dopachrome conversion. For decarboxylation, we used deprotonated carboxyl group (carboxylate ion) because the released CO_2 cannot be protonated. The calculated potential energy curves are shown in Fig. 2.5. A monotonically increasing profile was found for α-deprotonation. This indicates that α-deprotonation does not take place at this stage. β-Deprotonation showed 24.0 kcal/mol of the activation barrier, which is the lowest between the calculated three processes. Therefore, dopachrome conversion should start mainly from β-deprotonation. Although

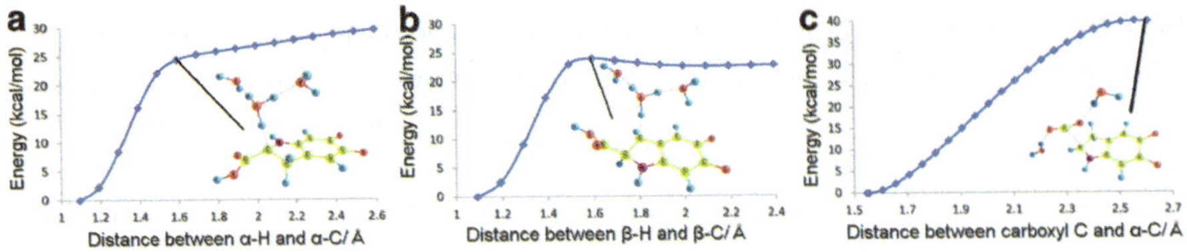

Fig. 2.5 Potential energy curves for **a** α-deprotonation, **b** β-deprotonation, and **c** decarboxylation of dopachrome (in the absence of Cu(II) coordination). Reprinted (with minor modification) in part from Ref. [21] with permission from Wiley

the calculated activation barrier for β-deprotonation is relatively high, there would also be other factors promoting this deprotonation in the actual system. For example, OH^- ions and buffer anions present at a low concentration may attack dopachrome, and then act as a proton acceptor instead of H_2O molecules.

Since the β-deprotonated structure is energetically unstable, this structure must be immediately reprotonated at different sites. As possible sites for the reprotonation, we considered 5-oxygen, 6-oxygen, and carboxylate group. Table 2.2 lists the calculated energetic preference for these reprotonated structures. We found that the O5-protonated structure was the most stable. To characterize the electronic state change by β-deprotonation, natural population analyses were performed. As shown in Fig. 2.6, 5-oxygen shows a considerably increased negative charge. This can be interpreted that the electron charge present in β-hydrogen was transferred to 5-oxygen

Table 2.2 Energetic stability of the dopachrome tautomers formed by proton rearrangement from β-carbon in the presence and absence of Cu(II) coordination at quinonoid group

Tautomer (Cu +/−)[a]	Reprotonation site[b]	Energy (kcal/mol)[c]	Gibbs free energy (kcal/mol)[d]
Initial structure (Cu−)	β-Carbon	0.0	0.0
A′ (Cu−)	Carboxyl	11.3	11.4
B′ (Cu−)	O5	−5.7	−5.1
C′ (Cu−)	O6	1.8	1.9
Initial structure (Cu+)	β-Carbon	0.0	0.0
A′ (Cu+)	Carboxyl	−12.5	−12.5
B′ (Cu+)	O5	3.3	2.7
C′ (Cu+)	O6	8.3	7.1

[a]Symbols for tautomers formed by proton rearrangement from β-carbon. The presence and absence of Cu(II) coordination at quinonoid group are, respectively, denoted as (Cu+) and (Cu−)
[b]Numbers in this column correspond to the labels in Fig. 2.1
[c]The energy origin was set to that of initial structure (Cu +/−)
[d]The energy origin was set to that of initial structure (Cu +/−). Temperature was set to 309.5 K as a condition of human body

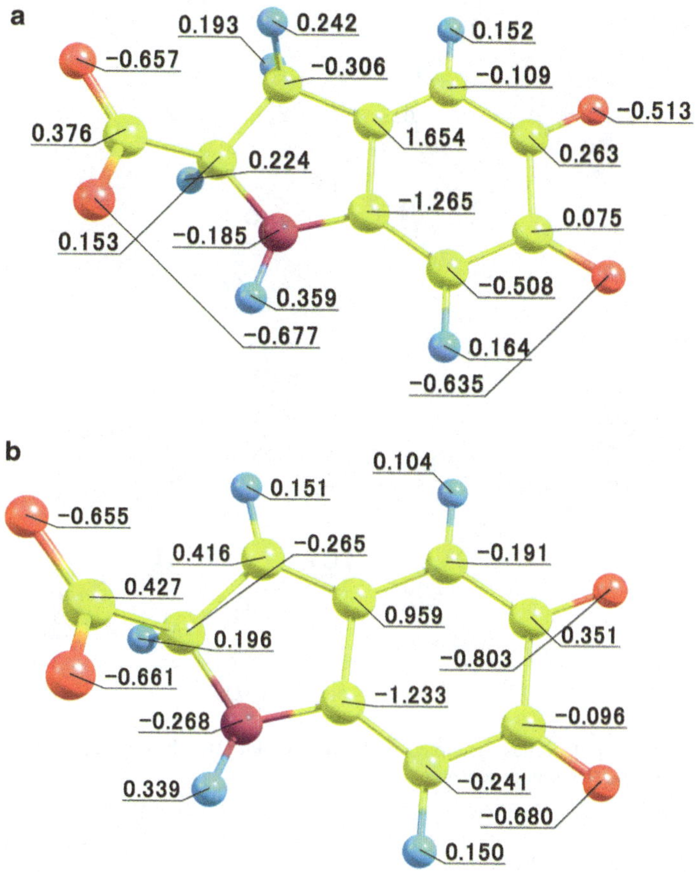

Fig. 2.6 Atomic charge (natural charge) distribution of dopachrome (**a**) before β-deprotonation and (**b**) after β-deprotonation. Elementary charge was used for the unit of charge

through the π-conjugated chain. This electronic structure presumably corresponds to the quinone methide intermediate identified by Sugumaran et al. [19, 20]. We also confirmed that α-deprotonation and decarboxylation proceeded with a similar charge transfer into 5,6-oxygens. As shown in the HOMO distribution (Fig. 2.7), the charge transfer during β-deprotonation mainly contributes to the occupation of the C − O antibonding orbital at 5-position.

Next, we considered the conversion processes from the obtained quinone methide intermediate (structure B′ defined in Table 2.2) to the possible products, namely DHI and DHICA. α-Deprotonation results in the formation of DHICA, while decarboxylation gives rise to the unprotonated DHI. Here, we calculated the activation barriers for the two processes. Figure 2.8 shows the obtained potential energy curves. As a result, we found that, from the quinone methide structure, α-deprotonation and decarboxylation requires 11.4 and 16.1 kcal/mol of activation energy, respectively.

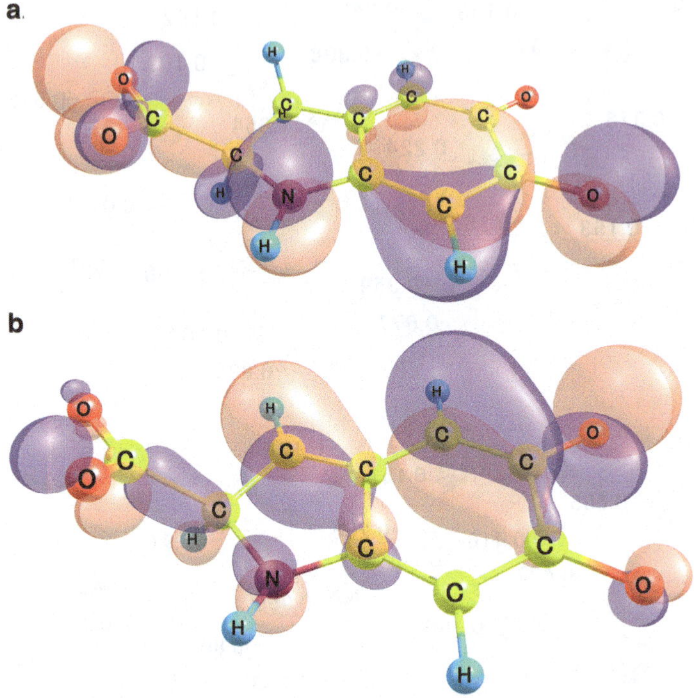

Fig. 2.7 Isosurfaces of highest occupied molecular orbital (HOMO) of dopachrome (**a**) before β-deprotonation and (**b**) after β-deprotonation

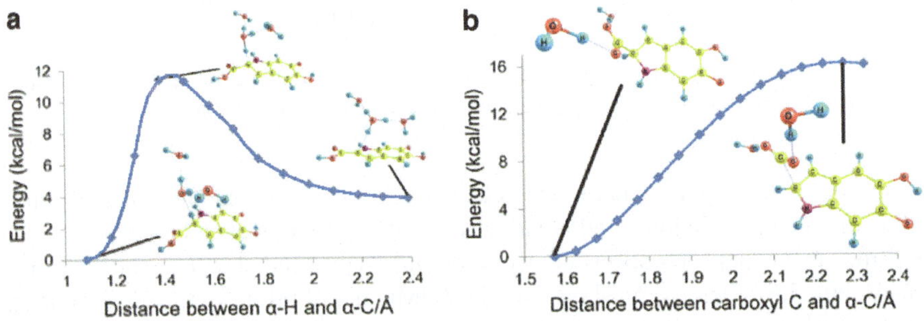

Fig. 2.8 Potential energy curves for **a** α-deprotonation and **b** decarboxylation of dopachrome conversion intermediate (where β-H is transferred to O5) (in the absence of Cu(II) coordination). Reprinted (with minor modification) in part from Ref. [21] with permission from Wiley

However, considering the final state of DHI, O6-protonation is also a necessary process. Therefore, as another possibility, we also calculated the energy profiles for α-deprotonation and decarboxylation after O6-protonation. As shown in the calculated potential energy curves (Fig. 2.9), the activation barrier for α-deprotonation and

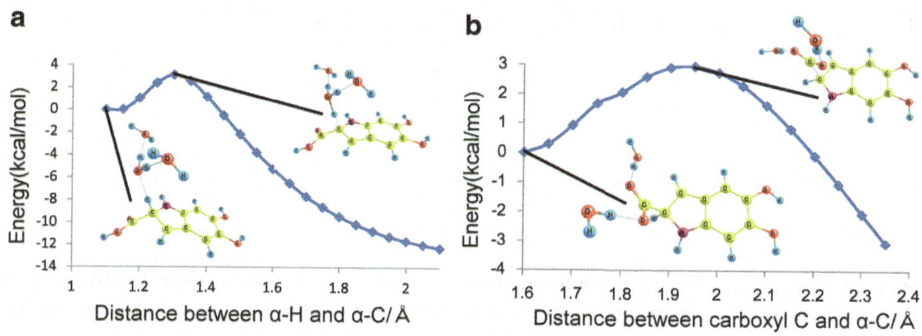

Fig. 2.9 Potential energy curves for **a** α-deprotonation and **b** decarboxylation of O6-protonated dopachrome conversion intermediate (where β-H is transferred to O5) (in the absence of Cu(II) coordination)

decarboxylation was 3.1 and 3.0 kcal/mol, respectively. Therefore, O6-protonation drastically promotes decarboxylation to produce DHI. Even though O6-protonation also decreased the activation barrier for α-deprotonation, this effect was weaker in comparison with that on decarboxylation. This is potentially due to the different nature of the two processes with respect to the concomitant charge transfer; decarboxylation results in negative charge transfer mainly to 6-oxygen, whereas α-deprotonation causes charge delocalization through the π-conjugated chain, which spreads also into carboxyl group. Although the difference in activation barrier is slight, our calculation clearly revealed the significance of O5- and O6-protonation for the selective formation of DHI. The base-catalyzed formation of DHICA previously reported is likely to correspond to a decreased O6-protonation rate and an increased α-deprotonation rate.

From this investigation, the mechanism of dopachrome conversion is proposed based on the activation barrier for various elementary steps. The initial step is β-deprotonation, followed by O5-protonation to produce a quinone methide intermediate. This intermediate further undergoes protonation at the remaining quinonoid oxygen, 6-oxygen. Finally, DHI is formed by decarboxylation. In this scheme, β-deprotonation showed the highest activation barrier, indicating that the subsequent proton rearrangement processes are not the rate-limiting step.

2.4 Dopachrome Conversion Mechanism with Cu(II) Coordination

Here, we consider dopachrome conversion with Cu(II) coordination at the quinonoid site of dopachrome. Figure 2.10 shows the calculated potential energy curves for α-deprotonation, β-deprotonation, and decarboxylation. Comparing the results in Figs. 2.5 and 2.10, Cu(II) coordination resulted in a significant decrease in the activation barrier for all the cases. With Cu(II) coordination, the activation barrier

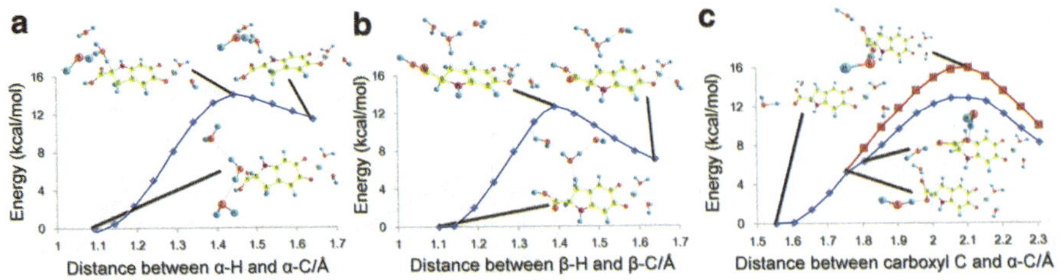

Fig. 2.10 Potential energy curves for **a** α-deprotonation, **b** β-deprotonation, and **c** decarboxylation of dopachrome (in the presence of Cu(II) coordination). Note that for decarboxylation, a dihedral angle along the dissociating C–C axis was intentionally fixed from the point where "sudden" rotation of carboxylate group occurred (see Sect. 2.2.). Diamonds denote the potential energies of fully optimized structures, while square boxes (in C) denote those of frozen dihedral angle along the dissociating C–C axis. Reprinted (with minor modification) from Ref. [22] with permission from Elsevier

for α-deprotonation, β-deprotonation, and decarboxylation was 14.0, 12.7, and 16.0 kcal/mol, respectively. Therefore, β-deprotonation is the most favorable process as in the Cu(II) absent case. Note that during the calculation for decarboxylation, the "immediate rotation" (mentioned in Sect. 2.2.) of dissociating carboxyl group occurred as shown in Fig. 2.10. Therefore, the corresponding dihedral angle was fixed from the point of immediate rotation.

As possible reprotonation sites after β-deprotonation, we considered 5-oxygen, 6-oxygen, and carboxylate group. Table 2.2 lists the energetic preference for these sites. In contrast to the Cu(II) absent case, where 5-oxygen was the most stable [B′ (Cu −)], the Cu(II) coordinated structure does not prefer the quinonoid sites for the reprotonation sites. Instead, the carboxylate group was the most stable in the presence of Cu(II) [A′ (Cu+)]. The instability of the O5- or O6-(re)protonated structure also manifested in spontaneous proton dissociation, which occurred when one H_2O molecule was placed near the protonated site with geometrical optimization. Therefore, the reprotonated states at the quinonoid sites are not only energetically less preferred but also impossible to be present in aqueous solutions.

In a similar manner to the Cu(II) absent case, a charge transfer into 5-oxygen was observed during β-deprotonation, while no significant changes in the charge state of copper was found. Thus, Cu(II) here functions as a Lewis acid but not as an oxidant. The decreased activation barriers for α-deprotonation, β-deprotonation, and decarboxylation are presumably due to the strong coordination bond of Cu(II), which prefers the localized negative charges at 5,6-oxygens resulting from these processes. Moreover, this strong coordination also does not prefer O5- and O6-protonation, which deprives the negative charge.

Next, we investigated the further processes to form DHI and DHICA from the proton-rearranged structure (A′ defined in Table 2.2). As shown in Table 2.2, the carboxyl group was significantly stabilized by the proton-capping so that the carboxylic acid dissociation is almost impossible. Accordingly, with Cu(II) coordination, decarboxylation is also an unfavorable process. Instead, α-deprotonation

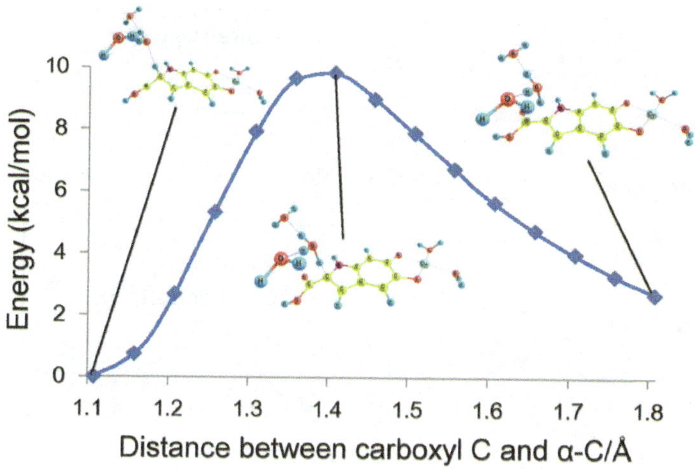

Fig. 2.11 Potential energy curves for α-deprotonation of dopachrome conversion intermediate (where β-H is transferred to carboxylate group) (in the presence of Cu(II) coordination). Reprinted (with minor modification) from Ref. [22] with permission from Elsevier

must take place to form DHICA. Figure 2.11 shows the potential energy curve for α-deprotonation from this stage. The calculated activation barrier was 9.8 kcal/mol, which is lower than that for the initial step β-deprotonation, indicating that this process is not the rate-determining step.

2.5 Proposed Scheme of Dopachrome Conversion

Figure 2.12 shows the proposed scheme of dopachrome conversion. Dopachrome conversion starts from β-deprotonation, and then in the absence of Cu(II) coordination, reprotonation occurs at 5-oxygen, while Cu(II) coordinated case prefers the carboxylate group for the reprotonation. Without Cu(II), DHI is formed by subsequent protonation at 6-oxygen, followed by decarboxylation. On the other hand, Cu(II) coordinated case has a stabilized carboxyl group by proton capping so that α-deprotonation alternatively takes place to form DHICA.

The rate-determining step is β-deprotonation, which is facilitated by basic pH and Cu(II) coordination. Although decarboxylation initially shows a high activation barrier, this is drastically decreased by protonation at 5,6-oxygens. It should be noted that the distantly located quinonoid group and carboxyl group are electronically connected. The reported selective formation of DHICA at basic pH and copper-catalyzed conditions may correspond to the inhibited protonation at 5,6-oxygens.

The proposed scheme is consistent with the reported experiments [10, 19, 20]. Our computational study explains the formation of the quinone methide intermediate and the selectively catalytic conversion in the presence of Cu(II) at the atomic level. Our

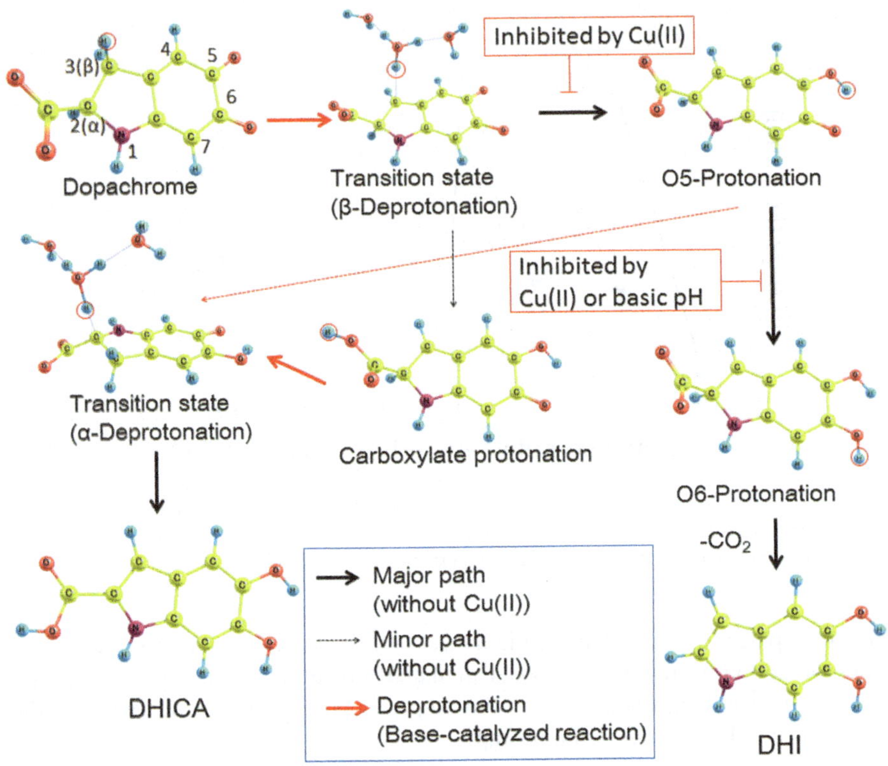

Fig. 2.12 Reaction scheme of dopachrome conversion

calculation results emphasize the importance of quinonoid protonation for promoting deprotonation and decarboxylation, where electronic manipulation of π-conjugated chains is the key to controlling the reaction.

References

1. J. Bustamante, L. Bredeston, G. Malanga, J. Mordoh, Role of melanin as a scavenger of active oxygen species. Pigment Cell Res. **6**, 348–353 (1993)
2. M. Tada, M. Kohno, Y. Niwano, Scavenging or quenching effect of melanin on superoxide anion and singlet oxygen. J. Clin. Biochem. Nutr. **46**, 224–228 (2010)
3. W. Korytowski, T. Sarna, Bleaching of melanin pigment: role of copper ions and hydrogen peroxide in autooxidation and photooxidation of synthetic dopa-melanin. J. Biol. Chem. **265**, 12410–12416 (1990)
4. L. Panzella, A. Napolitano, M. d'Ischia, Is DHICA the key to dopachrome tautomerase and melanocyte functions? Pigment Cell Melanoma Res. **24**, 248–249 (2010)

5. D. Kovacs, E. Flori, V. Maresca, M. Ottaviani, N. Aspite, M.L. Dell'Anna, L. Panzella, A. Napolitano, M. Picardo, M. d'Ischia, The eumelanin intermediate 5,6-dihydroxyindole-2-carboxylic acid is a messenger in the cross-talk among epidermal cells. J. Invest. Dermatol. **132**, 1196–1205 (2012)
6. P. Muneta, Enzymatic blackening in potatoes influence of pH on dopachrome oxidation. Am. Potato J. **54**, 387–393 (1977)
7. S. Ito, N. Suzuki, S. Takebayashi, S. Commo, K. Wakamatsu, Neutral pH and copper ions promote eumelanogenesis after the dopachrome stage. Pigment Cell Melanoma Res. **26**, 817–825 (2013)
8. H.S. Mason, The chemistry of melanin. III. mechanism of the oxidation of dihydroxyphenylalanine by tyrosinase. J. Biol. Chem. **172**, 83–99 (1948)
9. K. Wakamatsu, S. Ito, Preparation of eumelanin-related metabolites 5,6-dihydroxyindole 5,6-dihydroxyindole-2-carboxylic acid and their *O*-methyl derivatives. Anal. Biochem. **170**, 335–340 (1988)
10. A. Palumbo, M. d'Ischia, G. Misuraca, G. Prota, Effect of metal ions on the rearrangement of dopachrome. Biochim. Biophys. Acta **925**, 203–209 (1987)
11. A. Biesemeier, U. Schraermeyer, O. Eibl, Chemical composition of melanosomes lipofuscin and melanolipofuscin granules of human RPE tissues. Exp. Eye Res. **93**, 29–39 (2011)
12. Y. Liu, L. Hong, K. Wakamatsu, S. Ito, B. Adhyaru, C.Y. Cheng, C.R. Bowers, J.D. Simon, Comparison of structural and chemical properties of black and red human hair melanosomes. Photochem. Photobiol. **81**, 135–144 (2005)
13. A. Palumbo, F. Solano, G. Misuraca, P. Aroca, J.C. García-Borrón, J.A. Lozano, G. Prota, Comparative action of dopachrome tautomerase and metal ions on the rearrangement of dopachrome. Biophys. Acta **1115**, 1–5 (1991)
14. S. Commo, O. Gaillard, S. Thibaut, B.A. Bernard, Absence of TRP-2 in melanogenic melanocytes of human hair. Pigment Cell Res. **17**, 488–497 (2004)
15. S. Commo, K. Wakamatsu, I. Lozano, S. Panhard, G. Loussouarn, B.A. Bernard, S. Ito, Age-dependent changes in eumelanin composition in hairs of various ethnic origins. Int. J. Cosmet. Sci. **34**, 102–107 (2012)
16. S. Ito, Y. Nakanishi, R.K. Valenzuela, M.H. Brilliant, L. Kolbe, K. Wakamatsu, Usefulness of alkaline hydrogen peroxide oxidation to analyze eumelanin and pheomelanin in various tissue samples: application to chemical analysis of human hair melanins. Pigment Cell Melanoma Res. **24**, 605–613 (2011)
17. F. Solano, C. Jiménez-Cervantes, J.H. Martínez-Liarte, J.C. García-Borrón, J.R. Jara, J.A. Lozano, Molecular mechanism for catalysis by a new zinc-enzyme dopachrome tautomerase. Biochem. J. **313**, 447–453 (1996)
18. C.J. Vavricka, B.M. Christensen, J. Li, Melanization in living organisms: a perspective of species evolution. Protein Cell **1**, 830–841 (2010)
19. M. Sugumaran, H. Dali, V. Semensi, Formation of a stable quinone methide during tyrosinase-catalyzed oxidation of α-methyl dopa methyl ester and its implication in melanin biosynthesis. Bioorg. Chem. **18**, 144–153 (1990)
20. M. Sugumaran, V. Semensi, Quinone methide as a new intermediate in eumelanin biosynthesis. J. Biol. Chem. **266**, 6073–6078 (1991)
21. R. Kishida, Y. Ushijima, A.G. Saputro, H. Kasai, Effect of pH on elementary steps of dopachrome conversion from first-principles calculation. Pigment Cell Melanoma Res. **27**, 734–743 (2014)
22. R. Kishida, A.G. Saputro, H. Kasai, Mechanism of dopachrome tautomerization into 5,6-dihydroxyindole-2-carboxylic acid catalyzed by Cu(II) based on quantum chemical calculations. Biochim. Biophys. Acta **2015**, 281–286 (1850)
23. P. Hohenberg, W. Kohn, Inhomogeneous electron gas. Phys. Rev. **136**, B864–B871 (1964)
24. W. Kohn, L.J. Sham, Self-consistent equations including exchange and correlation effects. Phys. Rev. **140**, A1133–A1138 (1965)

25. M.J. Frisch, G.W. Trucks, H.B. Schlegel, G.E. Scuseria, M.A. Robb, J.R. Cheeseman, G. Scalmani, V. Barone, B. Mennucci, G.A. Petersson, H. Nakatsuji, M. Caricato, X. Li, H.P. Hratchian, A.F. Izmaylov, J. Bloino, G. Zheng, J.L. Sonnenberg, M. Hada, M. Ehara, K. Toyota, R. Fukuda, J. Hasegawa, M. Ishida, T. Nakajima, Y. Honda, O. Kitao, H. Nakai, T. Vreven, J.A. Montgomery, Jr., J.E. Peralta, F. Ogliaro, M. Bearpark, J.J. Heyd, E. Brothers, K.N. Kudin, V.N. Staroverov, R. Kobayashi, J. Normand, K. Raghavachari, A. Rendell, J. C. Burant, S.S. Iyengar, J. Tomasi, M. Cossi, N. Rega, J.M. Millam, M. Klene, J.E. Knox, J.B. Cross, V. Bakken, C. Adamo, J. Jaramillo, R. Gomperts, R.E. Stratmann, O. Yazyev, A.J. Austin, R. Cammi, C. Pomelli, J.W. Ochterski, R.L. Martin, K. Morokuma, V.G. Zakrzewski, G.A. Voth, P. Salvador, J.J. Dannenberg, S. Dapprich, A.D. Daniels, Ö. Farkas, J.B. Foresman, J.V. Ortiz, J. Cioslowski, D.J. Fox, Gaussian 09, Revision C. 01 (Gaussian, Inc., Wallingford CT, 2009)
26. A.D. Becke, Density-functional thermochemistry. III. The role of exact exchange. J. Chem. Phys. **98**, 5648–5652 (1993)
27. C. Lee, W. Yang, R.G. Parr, Development of the Colle-Salvetti correlation-energy formula into a functional of the electron density. Phys. Rev. B **37**, 785–789 (1988)
28. J.P. Foster, F. Weinhold, Natural hybrid orbitals. J. Am. Chem. Soc. **102**, 7211–7218 (1980)
29. J. Tomasi, B. Mennucci, R. Cammi, Quantum mechanical continuum solvation models. Chem. Rev. **105**, 2999–3093 (2005)
30. J.L. Pascual-Ahuir, E. Silla, and I. Tuñon, GEPOL: an improved description of molecular surfaces. III. A new algorithm for the computation of a solvent-excluding surface. J. Comput. Chem. **15**, 1127–1138 (1994)
31. A. Pasquarello, I. Petri, P.S. Salmon, O. Parisel, R. Car, É. Tóth, D.H. Powell, H.E. Fischer, L. Helm, A.E. Merbach, First solvation shell of the Cu(II) aqua ion: evidence for fivefold coordination. Science **291**, 856–859 (2001)
32. V.S. Bryantsev, M.S. Diallo, A.C.T. van Duin, W.A. Goddard III, Hydration of copper(II): new insights from density functional theory and the COSMO solvation model. J. Phys. Chem. A **112**, 9104–9112 (2008)
33. A.V. Marenich, C.J. Cramer, D.G. Truhlar, Universal solvation model based on solute electron density and on a continuum model of the solvent defined by the bulk dielectric constant and atomic surface tensions. J. Phys. Chem. **114**, 6378–6396 (2009)
34. W.A. Donald, E.R. Williams, An improved cluster pair correlation method for obtaining the absolute proton hydration energy and enthalpy evaluated with an expanded data set. J. Phys. Chem. B **114**, 13189–13200 (2010)
35. W.M. Haynes, *CRC Handbook of Chemistry and Physics*, 91st edn (2010–2011)

Chapter 3
Dopaquinone Conversion and Related Reactions

Abstract The biosynthetic pathway of melanin is branched into pheomelanogenesis and eumelanogenesis at the stage of dopaquinone conversion. In the presence of intracellular thiols such as cysteine, dopaquinone binds to the sulfhydryl group of thiols, whereas under lower concentration of thiols dopaquinone spontaneously undergoes intramolecular cyclization through the alanyl side chain. The binding of cysteine produces cysteinyldopa necessary for pheomelanogenesis, whereas the cyclized product, cyclodopa further transforms into eumelanin. In this chapter, we introduce computational studies for cyclization and thiol binding for dopaquinone and structurally similar *o*-quinones with the emphasis on the competitive behavior between the two reactions. As a result, remarkable charge redistributions were observed during cyclization and thiol binding. From this point of view, the HOMO and LUMO levels of *o*-quinone are pointed out as important factors affecting the reactivity and the competition between the two reactions. Furthermore, a mechanistic issue of thiol binding is also discussed based on the atomic-scale simulation results, which showed the presence of unusual reaction intermediate. Our approach clarifies branched reactions of dopaquinone and resembling *o*-quinones from atomic nuclei and electrons world.

Keywords Dopaquinone · *o*-quinone · Rhododendrolquinone · Cyclization · Thiol binding · Density functional theory

3.1 Introduction

3.1.1 Background—Competition Between Cyclization and Thiol Binding

In melanogenesis, the competitive reactions of dopaquinone controls the composition of the generated melanin. Known possible reactions of dopaquinone are cyclization and thiol binding (Fig. 3.1). The former results in eumelanogenesis while the latter corresponds to the initiation of pheomelanogenesis.

Dopaquinone includes two adjacent carbonyl groups in the benzene ring. Hence, this molecule is classified as *o*-quinone. In general, *o*-quinones are highly reactive

R. Kishida et al., *Melanin Chemistry Explored by Quantum Mechanics*,
https://doi.org/10.1007/978-981-16-1315-9_3

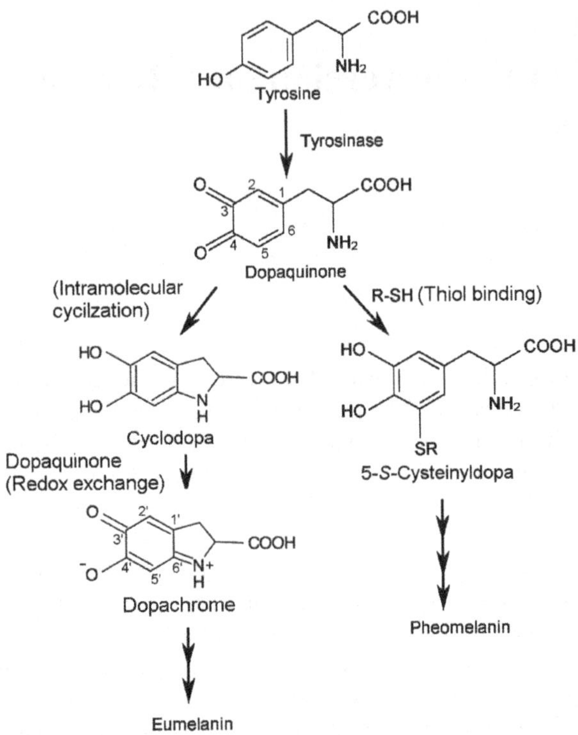

Fig. 3.1 Formation of dopaquinone and its subsequent conversions (cyclization and binding of thiols). Reprinted (with minor modification) from Ref. [20] with permission from Springer Nature

toward nucleophiles. When dopaquinone undergoes cyclization and thiol binding, an amino ($-NH_2$) group and a sulfhydryl ($-SH$) group acts as a nucleophile, respectively.

As introduced in Sect. 1.6, beyond the original extent of dopaquinone-based melanogenesis, reactions of structurally analogous *o*-quinones have been widely investigated. In particular, from a viewpoint of melanocyte-specific cytotoxicity, thiol binding is a clinically important target. Most of the tyrosinase substrates are *p*-substituted phenols or catechols. When these phenols or catechols are oxidized in the presence of tyrosinase, the corresponding *o*-quinones with the same *p*-substituent are formed (Fig. 3.2). This substituent controls the reactivity of *o*-quinones. For example, if the substituent is an amino- or hydroxyl-terminated hydrocarbon with chain length 2–4 carbon atoms, this molecule can be cyclized, giving rise to a five- to seven-membered ring [1].

The binding of cysteine containing a $-SH$ group with dopaquinone is a very rapid process, thus the competitive process of cyclization does not take place at cysteine concentration higher than 1 μmol/L [2]. From this viewpoint, pheomelanogenesis is considered to occur as the initial process, and then subsequently the generated eumelanins cover the pheomelanin particle after sufficient consumption of cysteine.

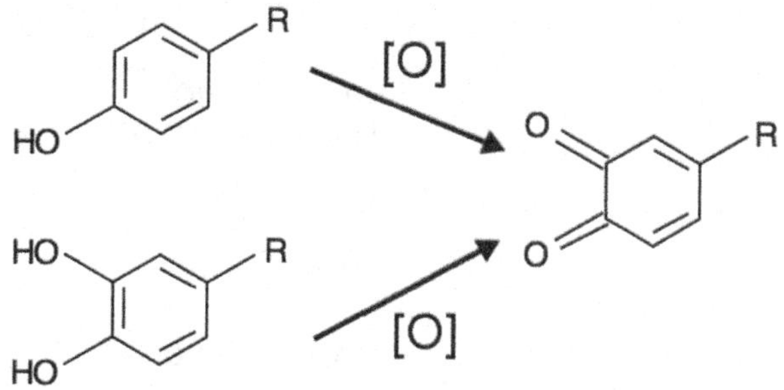

Fig. 3.2 Formation of *o*-quinones by tyrosinase-catalyzed oxidation of *p*-substituted phenols (upper left) or *p*-substituted catechols (lower left)

This model is called casing model. The validity of this model was confirmed by free electron laser-photoelectron emission microscopy (FEL-PEEM), in which a surface oxidation potential of a melanin sample (neuromelanin) was comparable with that of eumelanin [3, 4].

Dopaquinone is less reactive with bulky molecules like proteins [5]. This is presumably due to the large steric hindrance that prevents cysteine residues from binding with dopaquinone before cyclization. The competition between cyclization and thiol binding indicates that completion of one of the reaction makes dopaquinone less reactive to the other reaction. However, the *o*-quinone resulting from 4-S-CAP and RD-quinone can still undergo cyclization even after bounded with thiols [6, 7]. It can also be noted that cysteine binding sites are the 2,5-carbons of dopaquinone, which are different from the reaction site for cyclization, namely the 6-carbon. Therefore, there are no overlap of active sites for the two processes.

In a similar manner to dopaquinone, RD-quinone undergoes cyclization and thiol binding (Fig. 3.3). In addition to these two processes, another possible reaction is the addition of water (Fig. 3.3). This reaction proceeds at a slower rate than cyclization. However, the addition of water to dopaquinone has never been reported, probably due to the very rapid rate of cyclization. RD-quinone has a hydroxyl group and a methyl group at the end of the side chain. When RD-quinone cyclizes, this hydroxyl group forms a covalent bond with a benzene ring carbon (C6). This 6-carbon is also an active site for the addition of water, providing a competition. Although RD-quinone is converted to catechols by cyclization and the addition of water (Fig. 3.3), the resulting catechols are immediately oxidized to form RD-cyclic quinone and RD-*p*-hydroxy-quinone, respectively.

3.1.2 Background—Cyclization Kinetics for o-quinones

The cyclization rate of an *o*-quinone reflects its inherent reactivity. Dopaquinone is an α-amino acid whose structure is derived from the uncarboxylated basic structure dopamine quinone (an oxidized form of the neurotransmitter dopamine). Kinetic influence of introducing various substituents into dopamine quinone has

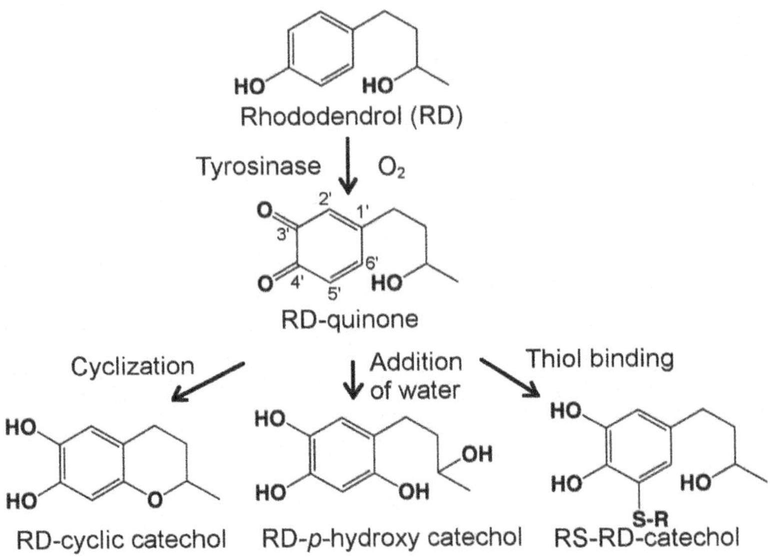

Fig. 3.3 Formation of rhododendrol-quinone (RD-quinone) and its subsequent conversions (cyclization, binding of thiols, and addition of water). For RD-quinone, binding of cysteine occurs at 5′-carbon, whereas in the case of RD-cyclic quinone the same reaction occurs at 2′-carbon instead. RD-cyclic quinone and RD-*p*-hydroxycatechol exist in a chemical equilibrium, where RD-cyclic quinone can convert to RD-*p*-hydroxycatechol via hemiacetal intermediate in the presence of acid catalysts [6]

been widely investigated and reviewed by Land et al. [1]. Especially, introducing carboxyl group (i.e. dopaquinone) and *N*-alkyl groups promoted cyclization [1, 4, 8, 9]. In contrast, a drastic decrease in cyclization rate was observed when *N*-acyl groups were introduced [10].

Due to the competitive reactions, cyclization and thiol binding, the yield of the thiol-bound product also depends on the cyclization rate even at the same thiol concentration. From an experiment that investigated the binding of bovine serum albumin (BSA) with various *o*-quinones, it was found that dopamine quinone binds BSA in a yield higher than the case of dopaquinone, and epinephrine binds BSA in a yield lower than the case of the non *N*-methylated analog norepinephrine. In other words, the presence of α-carboxyl group and *N*-alkyl group lowers the *o*-quinone's reactivity to thiols. This reduced thiol binding corresponds to the accelerated cyclization.

Since *o*-quinoneamines are basic compounds, most of the amino groups are present in the protonated form at neutral pH, where the bonding sites are fully occupied. To newly form a covalent bond, the amino groups thus need to dissociate a proton at first. Kinetic studies using pulse radiolysis have pointed out that the deprotonation and the reprotonation (backward process) can be in a quasi-equilibrium due to the much slower subsequent process, namely nucleophilic amino attack to complete cyclization [1, 11–13]. This quasi-equilibrium manner results in the rate of the overall reaction which is proportional to the acidity constant of amino group and to the rate constant for the nucleophilic addition by amino group. Therefore, the overall reaction rate depends on the two factors: basicity and nucleophilicity of the side chain. Especially, side chain nucleophilicity is not directly amenable to an experimental measurement. Thus, the relation between the side chain structure and the nucleophilicity for cyclization remains to be explored.

3.1.3 Background—Binding of Cysteine with Dopaquinone

Cysteine forms a covalent bond with dopaquinone at 5-carbon or 2-carbon, but not at 6-carbon. In contrast, cyclization occurs at 6-carbon. Although cyclization at 2-carbon may also be theoretically possible, a previous computational investigation excluded this possibility [1]. This may be due to the intrinsic energy difference between before and after the cyclic bond formation, which perturbs the π-conjugated chain. Therefore, it is straightforward to consider that this preference of 6-carbon over 2-carbon is found for general nucleophiles. However, in the case of the cysteine binding, the yield of 6-adduct reported is only 1%, and the major products were 5-adduct (74%) and 2-adduct (14%) [2, 14–16]. Note that dicysteinyldopa, where 5-carbon and 2-carbon are both bound, was also reported with a 5% yield [2, 15, 16].

As a mechanism for the initial step of the cysteine binding, 1,6-Michael addition mechanism has been proposed [17–19]. In this mechanism, the sulfhydryl group in cysteine forms a covalent bond at 5-carbon (or 2-carbon), and then 3-oxygen (or 4-oxygen) is protonated. The rate for the cysteine binding increases with the cysteine concentration. Although the binding rate linearly increases at lower concentration of cysteine, this increase becomes more gradual as the concentration gets higher [19]. This behavior indicates the presence of a reaction intermediate of the cysteine binding, and thus supports the 1,6-Michael addition mechanism.

Furthermore, the cysteine binding rate is positively correlated with pH [18, 19]. Therefore, the binding of cysteine must be initiated by sulfhydryl deprotonation. Jameson et al. proposed a kinetic model based on the 1,6-Michael addition mechanism, and compared the binding of the amino-free cysteine analog thioglycolic acid with that of cysteine [19]. As a result, the intermediate structure resulting from the binding of cysteine was unstable as compared to the case of the binding of thioglycolic acid. Therefore, the amino group in cysteine would have an important role in the reaction with dopaquinone.

3.1.4 Theoretical Approach for *o*-Quinone Reactions

In this chapter, we describe our theoretical investigations, which were conducted to understand the two reactions involving dopaquinone and similar *o*-quinones, namely cyclization and binding of thiols [20–22]. As a computational method, we employed density functional theory-based first principles calculation, with parameters described in Sect. 3.2.

Section 3.3 describes the simulated results for the competition between cyclization and thiol binding of dopaquinone and RD-quinone. As a model of thiols, the simplest structure methane thiolate ion (SCH_3^-) was chosen. We compared the binding energy of SCH_3^- to dopaquinone and RD-quinone before and after the cyclization. As the cyclized structure of dopaquinone and RD-quinone, we used dopachrome and RD-cyclic quinone, respectively. As a result, dopaquinone case showed an unstable SCH_3^--bound state after cyclization, while RD-quinone could bind SCH_3^- even after cyclization.

Section 3.4 describes computational studies on the initial cyclization process for various *o*-quinones. As an elementary step of the initial cyclization process, we considered the C6–N or C6–O cyclic bond formation. For the cyclization of *o*-quinoneamines, we chose dopaminequinone and its analogs, namely α-carboxylated, *N*-methylated, and *N*-formylated derivatives. Besides, the methylene- ($-CH_2-$) inserted analogs in their side chain were also considered. Furthermore, as an example of cyclization with hydroxyl group, RD-quinone cyclization was investigated. As a result, the α-carboxylated and *N*-methylated dopaminequinone showed decreased activation barriers for the C6–N cyclic bond formation. In contrast, introduction of *N*-formyl group resulted in a remarkable increase in the activation barrier. Using the methylene-inserted structures, the increased hydrocarbon side chains showed slightly decreased activation barriers. In the case of RD-quinone, the C6–O cyclic bond formation was not possible from the electroneutral structure, indicating the necessity of hydroxyl deprotonation as the initial step.

Section 3.5 describes a mechanistic investigation on binding of cysteine with dopaquinone. We found that cysteine thiolate could form a bound state on C5, C2, C6, C3–C4 bridge, and C1 of dopaquinone. Especially, we identified C3–C4 bridge site, which was not experimentally found, and the binding on C3–C4 bridge showed the highest binding energy among all the sites, including C5 and C2. Furthermore, the binding energy at C2 was not higher than that at C6. Therefore, the reported preference of C2 over C6 for the thiol binding site cannot be explained by the energetic stability. From these results, we proposed the binding mechanism, in which cysteine approaches C3–C4 bridge, and then migrates to C5 or C2, followed by proton rearrangements to give 5-*S*-cysteinyldopa or 2-*S*-cysteinyldopa, respectively.

3.2 Calculation Methods for Simulating Dopaquinone Conversion

As in the previous chapter, we conducted density functional theory-based first principles calculation [23, 24] with the Becke's three-parameter hybrid functional [25] combined with the Lee-Yang-Parr correlation functionals (B3LYP) [26]. Calculations were carried out with 6-31 ++G(d,p) basis set using the Gaussian09 computational package [27]. The atomic charges were estimated by the natural atomic orbital analysis [28]. Considering aqueous phase reactions, we used PCM to describe the solvent–solute interaction [29, 30].

As a descriptor of nucleophilicity, the condensed-to-atom Fukui indices (the derivatives of the electron density with respect to the total number of electrons, namely the normalized local softness of the electronic system) were calculated using the finite difference approximation [31, 32]. In this approximation, the $N + 1$ and $N - 1$ electron systems were calculated using the structure of the N electron system, and then the atomic charges were determined by the natural population orbital analysis.

To obtain the activation barriers, one-dimensionally projected potential energy curves were calculated along the direction in which the bond length increases with a step size increment of 0.05 or 0.10 Å. During the calculations, all the degrees of freedom except for the one specifically chosen to be frozen (as specified by the structure of the reaction) were allowed to relax. To find the transition states of the reactions, we used the synchronous transit and quasi-Newton (STQN) method [33]. Note that the transition state structures and the activation energies obtained from the calculated potential energy curves and from the STQN method were almost identical.

3.3 Competition Between Cyclization and Thiol Binding—Comparison Between Dopaquinone and Rhododendrol Quinone

As described in Sect. 3.1, *o*-quinones competitively undergo cyclization and thiol binding even though their active sites do not overlap. Here, we considered metastable species before and after cyclization to discuss the change of the binding ability to thiols. As a model of thiols, the simplest structure methane thiolate ion (SCH_3^-) was chosen. For the comparison of *o*-quinone structures, we chose dopaquinone and RD-quinone.

First, we investigated the change in the electronic state and the total energy when SCH_3^- ion binds with dopaquinone. As shown in Fig. 3.4, through the C–S bond formation, a charge transfer of approximately one electron occurred into dopaquinone, which occupied the LUMO of dopaquinone. With this, the change in the total energy of this bond formation would be mainly determined by the LUMO level of dopaquinone. The binding energy is calculated as

$$E_{\mathrm{B}} = (E_{\mathrm{SCH_3^-}} + E_{\mathrm{dopaquinone}}) - E_{\mathrm{SCH_3^-}\,+\mathrm{dopaquinone}}$$

where $E_{\mathrm{SCH_3^-}}$ is the energy of isolated SCH_3^-, $E_{\mathrm{dopaquinone}}$ is the energy of isolated dopaquinone, and $E_{\mathrm{SCH_3^-}\,+\mathrm{dopaquinone}}$ SCH_3^- interacting with dopaquinone. In this case, the value of the binding energy was 4.38 kcal/mol. The positive value of the binding energy means a stable bound state.

Next, we investigated the change in the electronic state and the total energy by the SCH_3^--binding after cyclization. As a metastable cyclized product, we used the

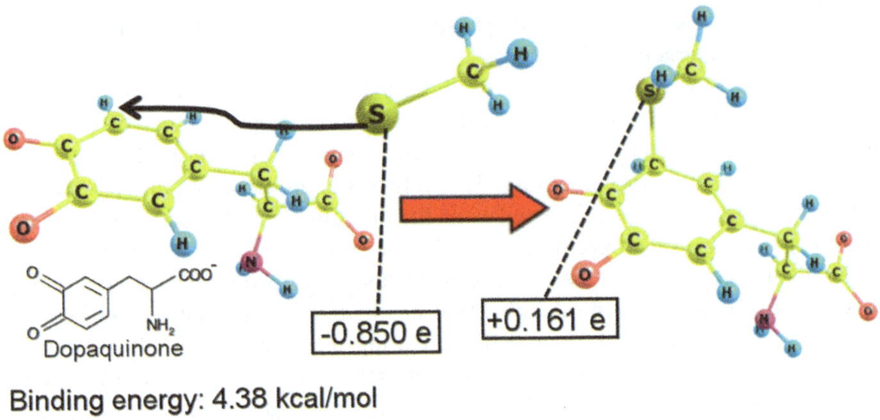

Fig. 3.4 Binding of methane thiolate ion SCH_3^- on dopaquinone (before cyclization) and corresponding charge re-distribution. Reprinted (with minor modification) from Ref. [20] with permission from Springer Nature

oxidized form of cyclodopa, namely dopachrome. As in the case before cyclization, a charge transfer of approximately one electron occurred into dopachrome (Fig. 3.5). For this interaction, the binding energy was −10.38 kcal/mol. This negative binding energy corresponds to the instability of the SCH_3^--bound state (the thiolate-bound state is less stable than the isolated state.). Therefore, the binding of SCH_3^- ion to the cyclized dopaquinone (dopachrome) is not energetically preferred. Note that the thiolate-bound structure that was computationally obtained is located at a local and not the global minimum of the potential energy surface. Due to cyclization, the LUMO level of the dopachrome was up-shifted by 0.66 eV (15.22 kcal/mol) with respect to the uncyclized dopaquinone. In contrast, the HOMO level was slightly

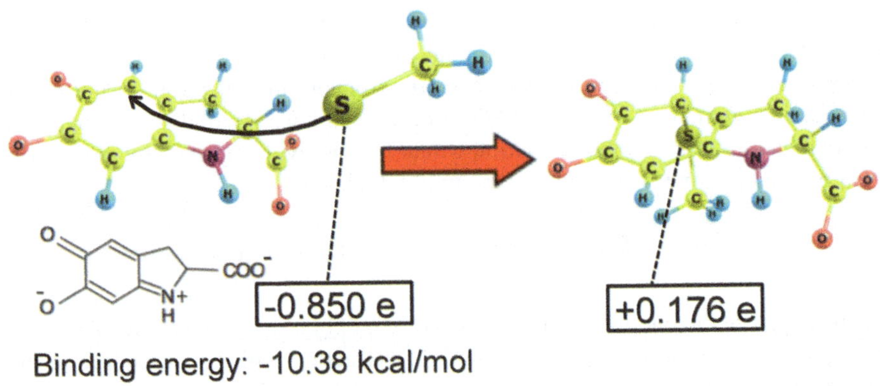

Fig. 3.5 Binding of methane thiolate ion SCH_3^- on dopachrome (after cyclization of dopaquinone) and corresponding charge re-distribution. Reprinted (with minor modification) from Ref. [20] with permission from Springer Nature

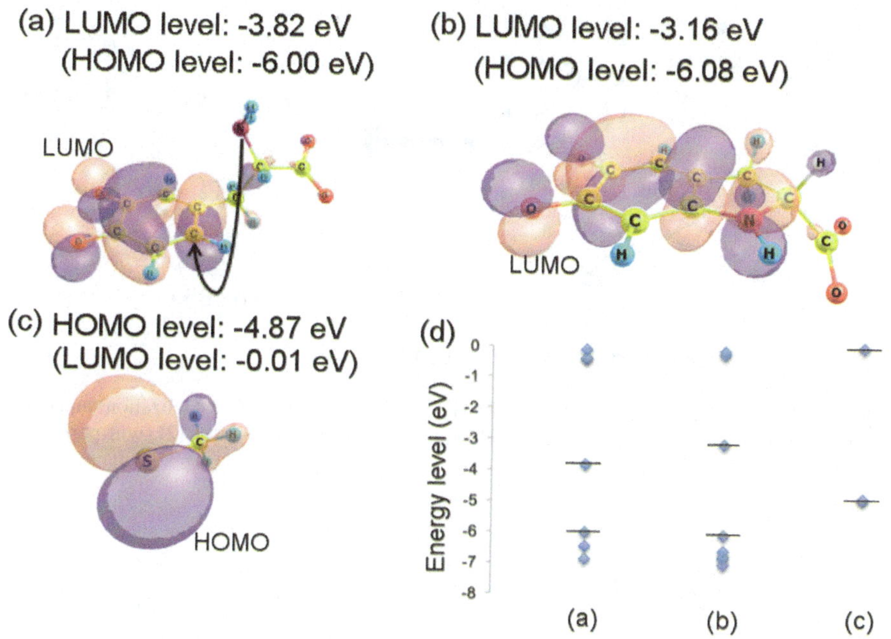

Fig. 3.6 Isosurface plots for LUMOs of **a** dopaquinone and **b** dopachrome, and HOMO of **c** methane thiolate ion SCH_3^-. Energy level diagram is shown in **d**, where horizontal lines show HOMO and LUMO. The vacuum level was chosen as the origin of energy level. Reprinted (with minor modification) from Ref. [20] with permission from Springer Nature

up-shifted by only 0.08 eV (1.84 kcal/mol). The electronic energy levels before and after the dopaquinone cyclization are summarized in Fig. 3.6.

For comparison, we investigated the change in the electronic state and the total energy of RD-quinone upon the SCH_3^--binding. As shown in Fig. 3.7, the thiol binding resulted in a charge transfer of approximately one electron into RD-quinone. As in the case of dopaquinone, this charge transfer occurred accompanying an electron occupation of the LUMO of RD-quinone. The binding energy was 7.61 kcal/mol, the positive sign means a stable binding of SCH_3^-.

Furthermore, we investigated the SCH_3^--induced change in the electronic state and the total energy of a cyclized RD-quinone, namely RD-cyclic quinone, which is the oxidized form of RD-cyclic catechol. In a similar manner to the case of the uncyclized RD-quinone, the sulfur atomic charge of approximately one electron was transferred to RD-cyclic quinone (Fig. 3.8). The binding energy was 4.15 kcal/mol. The positive value indicates that RD-quinone can bind thiolates even after cyclization unlike the case of dopaquinone. After cyclization, the LUMO level of RD-cyclic quinone was up-shifted by 0.23 eV (5.24 kcal/mol) from that of the uncyclized RD-quinone. Unlike the case of dopaquinone, the HOMO level was also remarkably up-shifted by 0.49 eV (11.25 kcal/mol). The electronic energy levels before and after the RD-quinone cyclization are summarized in Fig. 3.9.

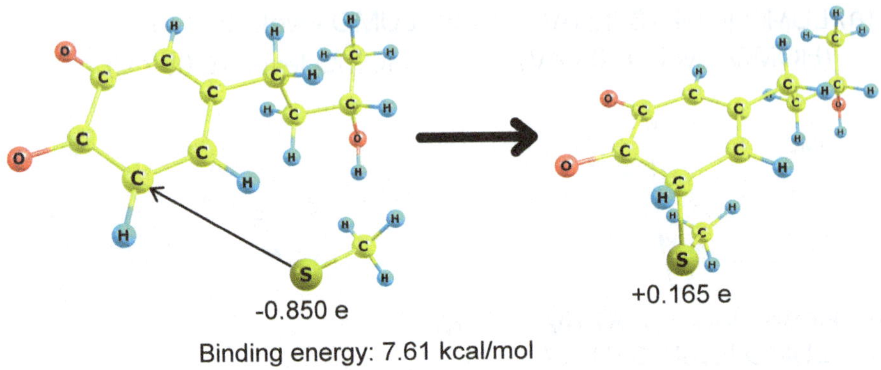

Fig. 3.7 Binding of a methane thiolate ion SCH_3^- on RD-quinone (before cyclization) and corresponding charge re-distribution. Reprinted (with minor modification) from Ref. [21] with permission from Physical Society of Japan

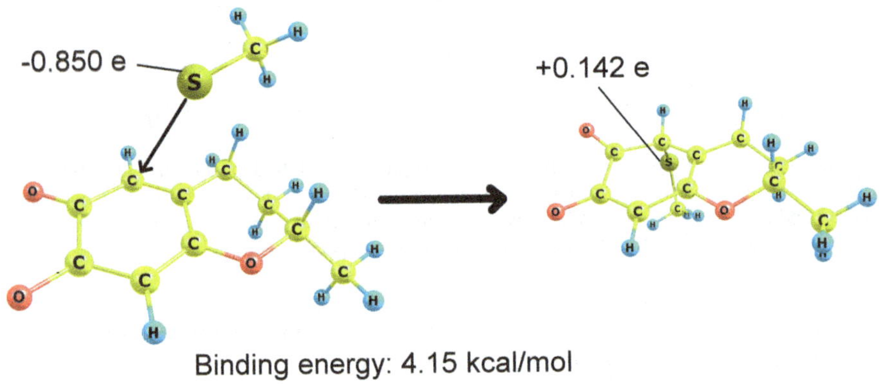

Fig. 3.8 Binding of a methane thiolate ion SCH_3^- on RD-cyclic quinone (after cyclization of RD-quinone) and corresponding charge re-distribution. Reprinted (with minor modification) from Ref. [21] with permission from Physical Society of Japan

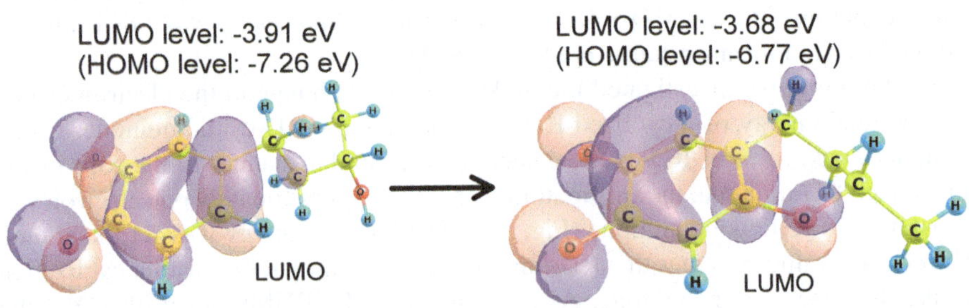

Fig. 3.9 Isosurface plots for LUMOs of **a** RD-quinone and **b** RD-cyclic quinone with change in energy level. The vacuum level was chosen as the origin of energy level. Reprinted (with minor modification) from Ref. [21] with permission from Physical Society of Japan

o-Quinones become less reactive to thiols after cyclization. The influence of cyclization on the reactivity to thiols was different between dopaquinone and RD-quinone. In this sense, RD-quinone can be said to be oriented to thiol binding more than dopaquinone. After dopaquinone forms the C6–N cyclic bond, the spatial distribution of the LUMO around the C6–N shows a nodal plane through the middle of the bond axis, indicating an antibonding orbital interaction between the lone pair orbital of the amino group (mainly appeared as the HOMO) and one of the π-orbital of the benzene ring (appeared as the LUMO) [Fig. 3.6b]. In the case of the C6–O cyclic bond formation of RD-quinone, the LUMO distribution also exhibits an antibonding orbital interaction between the lone pair orbital of the hydroxyl group (mainly appeared as the HOMO) and one of the π-orbital of the benzene ring (as the LUMO) [Fig. 3.9b]. RD-quinone cyclization proceeds by forming an ionic C–O bonding, which is less covalent character compared to that of the C–N bonding. This gave the larger HOMO–LUMO gap (3.35 eV) of RD-quinone than that (2.18 eV) of dopaquinone. As can be theoretically predicted (for typical example, by the second-order perturbation theory), an orbital interaction gives an energy shift, which becomes large when the energy levels of the two orbitals to be interacted are closer as in the cases of covalent bonding. Accordingly, the larger LUMO shift (toward the vacuum level) in the RD-quinone cyclization can be explained based on the ionic character of the C–O bonding.

As mentioned above, the binding energy of thiols would be explained based on the LUMO level of *o*-quinones. Besides the case of dopaquinone and RD-quinone, we further extended this investigation into more general *o*-quinones. Figure 3.10

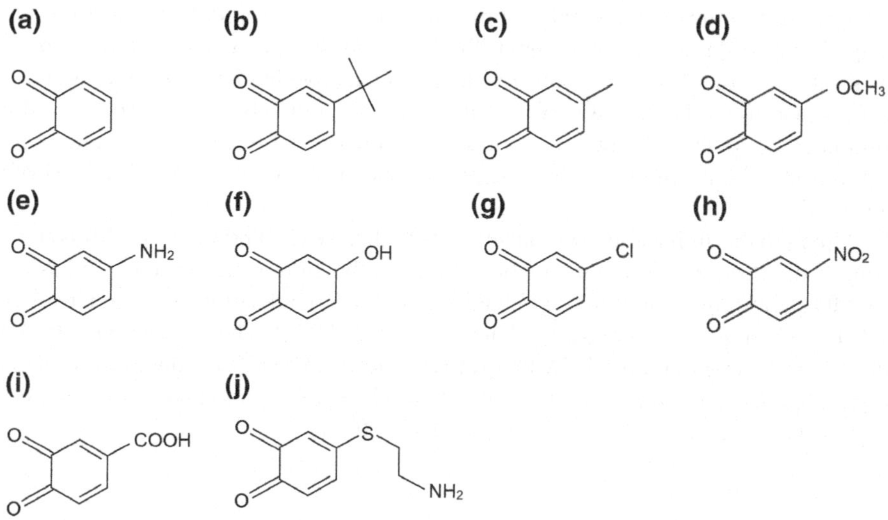

Fig. 3.10 *o*-Quinone structures. **a** 1,2-Benzoquinone, **b** 4-*tert*-Butyl-1,2-benzoquinone, **c** 4-Methyl-1,2-benzoquinone, **d** 4-Methoxy-1,2-benzoquinone, **e** 4-Amino-1,2-benzoquinone, **f** 4-Hydroxy-1,2-benzoquinone, **g** 4-Chloro-1,2-benzoquinone, **h** 4-Nitro-1,2-benzoquinone, **i** 4-Carboxy-1,2-benzoquinone, **j** 4-*S*-Cysteaminyl-1,2-benzoquinone

Table 3.1 Binding energies of methane thiolate ion to various *o*-quinones

Label[a]	*o*-Quinone	Binding energy E_B (kcal/mol)[b]	LUMO level (eV)[c]
A	1,2-Benzoquinone	10.0	−4.0
B	4-*tert*-Butyl-1,2-benzoquinone	6.6	−3.9
C	4-Methyl-1,2-benzoquinone	8.3	−3.9
D	4-Methoxy-1,2-benzoquinone	2.1	−3.7
E	4-Amino-1,2-benzoquinone	−0.4	−3.5
F	4-Hydroxyl-1,2-benzoquinone	6.0	−3.8
G	4-Chloro-1,2-benzoquinone	14.1	−4.2
H	4-Nitro-1,2-benzoquinone	21.5	−4.7
I	4-Carboxyl-1,2-benzoquinone	15.9	−4.4
J	4-*S*-Cysteaminyl-1,2-benzoquinone	8.8	−3.9
K	Dopaquinone	4.5	−3.8
L	Rhododendrol quinone	7.6	−3.9

[a]The structures of compounds labeled as A–J are shown in Fig. 3.10

[b]$E_B = (E_{SCH_3^-} + E_{dopaquinone}) - E_{SCH_3^- + dopaquinone}$

[c]The energy origin was set to the vacuum level

shows the structural formula of the calculated *o*-quinones. These have simple *o*-quinone structures, and include both electron-donating groups (functional groups that release its electron density to the neighboring systems in the reactions) and electron-withdrawing groups (functional groups that pull the electron density out from the neighboring systems in the reactions). Table 3.1 lists the calculated results. In particular, introduction of amino groups resulted in a negative binding energy of SCH_3^-. Correspondingly, the LUMO level of these amino acid substituent also showed a significantly up-shifted value. In contrast, introduction of nitro group resulted in a significant increase in the binding energy consistent with the down-shifted LUMO level.

Although the amino and nitro substituents both have C–N bond, the LUMOs show a markedly different orbital interaction. Namely, the amino-substituted case has the LUMO exhibiting an antibonding behavior at the C–N bond region, whereas the nitro-substituted case shows a bonding character of the LUMO. The correlation between the binding energy and the LUMO level was shown in Fig. 3.11, demonstrating an almost linear correlation. Therefore, the LUMO level of *o*-quinones can be generally regarded as a descriptor of the binding ability to thiols.

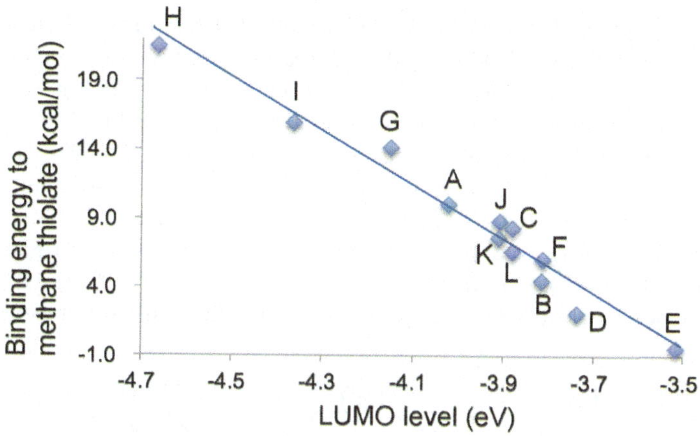

Fig. 3.11 Correlation between binding energy of thiols on *o*-quinones and LUMO levels

3.4 Cyclization of Dopamine Quinone Analogs

To describe *o*-quinone cyclization, we investigated the C6–N cyclic bond formation of *o*-quinoneamines and the C6–O cyclic bond formation of RD-quinone. Specifically, as cyclizable *o*-quinoneamines, we chose (a) dopamine quinone and its analogs, namely (b) dopaquinone, (c) *N*-methyl-dopaminequinone, (d) *N*-formyl-dopaminequinone, and [(a′)–(d′)] the methylene- ($-CH_2-$) inserted analogs in their side chain (Fig. 3.12). (a)–(d) correspond to five-membered ring formation, while (a′)–(d′) give rise to six-membered rings. Note that *N*-methyl and *N*-formyl substituent is an electron-donating and an electron-withdrawing group, respectively.

The *o*-quinoneamines have a hydrocarbon side chain which is flexible in rotation around C–C or C–N axis. To find typical conformational features, we calculated the potential energy surface along the side chain rotation of (a) dopaminequinone as a representative case. The conformation of the side chain is specified by two dihedral

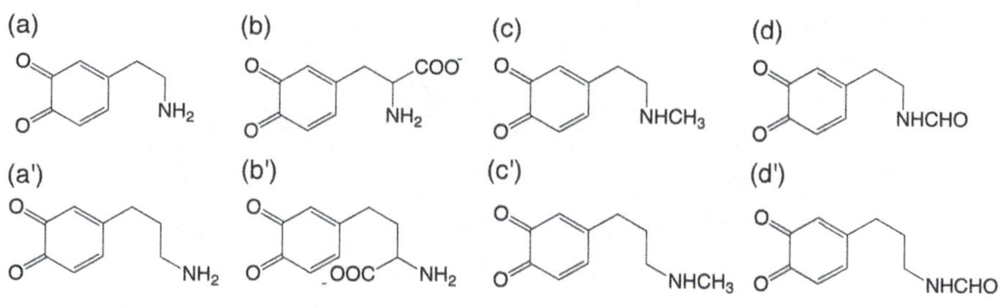

Fig. 3.12 *o*-Quinone structures. **a** Dopaminequinone, **b** Dopaquinone, **c** *N*-Methyl-dopaminequinone, **d** *N*-Formyl-dopaminequinone, **a′** *Homo*-dopaminequinone, **b′** *Homo*-dopaquinone, **c′** *N*-Methyl-*homo*-dopaminequinone, **d′** *N*-Formyl-*homo*-dopaminequinone. Numbering denoted in **a** is based on common nomenclature

angles, θ_1 and θ_2 as defined in Fig. 3.13. [Note that, for the methylene-inserted cases, three angles are needed to specify the rotation. See Fig. 3.13a′.] The obtained potential energy surface along the side chain rotation is shown in Fig. 3.14. As can be seen in the narrower contour, the rotation along θ_1 (around C–C axis) requires higher activation energy than that along θ_2 (around C–N axis). The activation barrier for the rotation along θ_2 is very low (less than 50 meV), indicating that the amino group can be almost freely twisting.

The potential energy curves along the C6–N cyclic bond formation of *o*-quinoneamines [(a)–(d), (a′)–(d′)] are shown in Figs. 3.15 and 3.16. The C6–N interatomic distance must decrease along the reaction path. Thus, we chose the C6–N

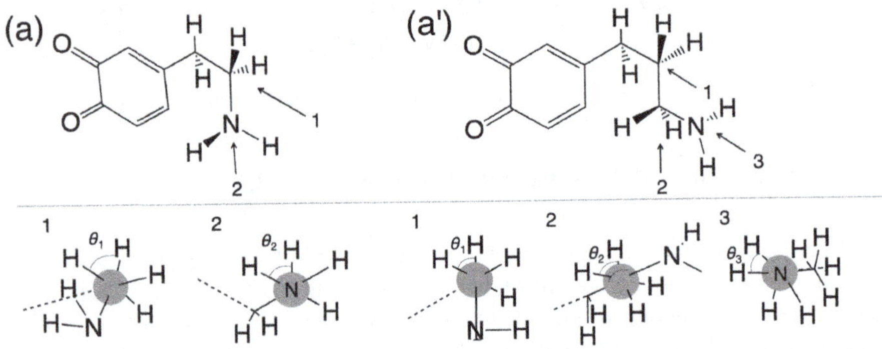

Fig. 3.13 Conformation of the hydrocarbon side chain in **a** dopaminequinone and **a′** *homo*-dopaminequinone. The arrows in the upper half roughly represent the view angles in the lower half. Definition of dihedral angles (θ_1, θ_2, and θ_3) was shown in the lower half

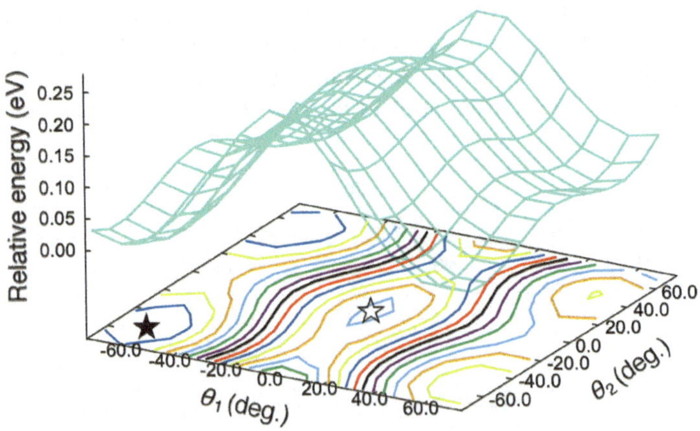

Fig. 3.14 Potential energy surface for two dihedral angles (θ_1 and θ_2, defined in Fig. 3.13) of **a** dopaminequinone. Black and white stars correspond to the most stable and the eclipsed conformation, respectively. The contour was plotted by 10 meV spacing. Note that all molecular degrees of freedom except for the two dihedral angles were allowed to relax

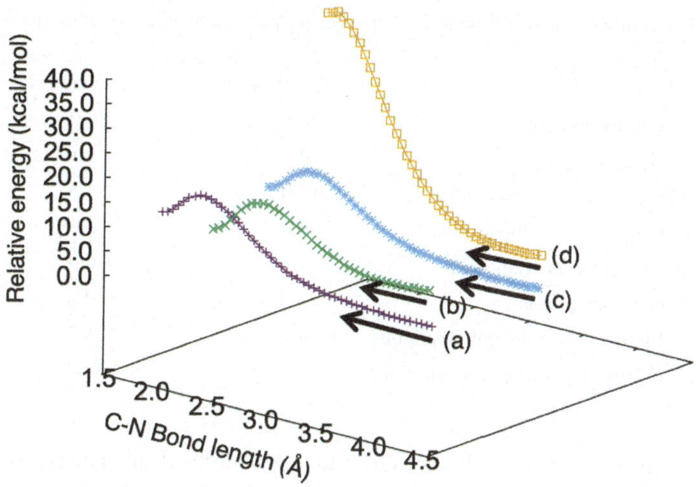

Fig. 3.15 Potential energy curves for **a** dopaminequinone, **b** dopaquinone, **c** *N*-methyl-dopaminequinone, and **d** *N*-formyl-dopaminequinone along C6–N bond formation. Each energy origin was chosen as that of each corresponding initial state (energy minimum)

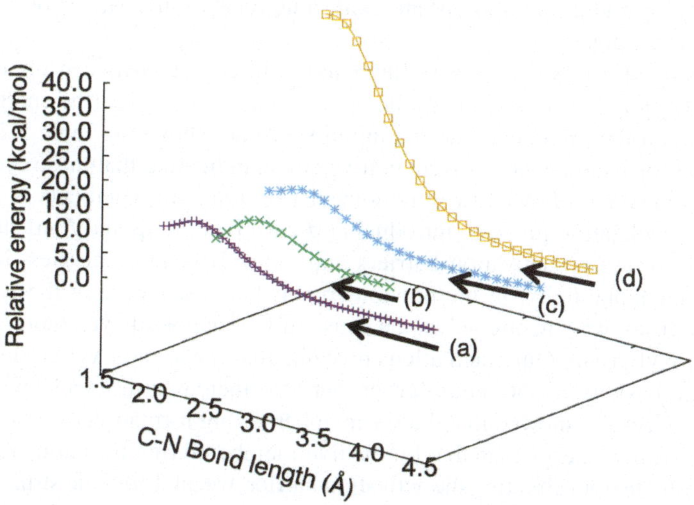

Fig. 3.16 Potential energy curves for **a** *homo*-dopaminequinone, **b** *homo*-dopaquinone, **c** *N*-methyl-*homo*-dopaminequinone, and **d** *N*-formyl-*homo*-dopaminequinone along C6–N bond formation. Each energy origin was chosen as that of each corresponding initial state (energy minimum)

Table 3.2 Activation barriers and bond formation energies for C6–N cyclic bonding

Label	*o*-Quinone	Activation barrier (kcal/mol)	Bond formation energy (kcal/mol)
(a)	Dopaminequinone	14.8	9.0
(b)	Dopaquinone	9.7	2.1
(c)	*N*-Methyl-dopaminequinone	12.0	5.5
(d)	*N*-Formyl-dopaminequinone	39.0	38.3
(a′)	*Homo*-dopaminequinone	9.5	6.9
(b′)	*Homo*-dopaquinone	6.2	0.0
(c′)	*N*-Methyl-*homo*-dopaminequinone	7.6	5.5
(d′)	*N*-Formyl-*homo*-dopaminequinone	38.1	37.8

distance as the reaction coordinate. Note that we allowed all degrees of freedom except for the relative coordinate between C6 and the amino N to relax during the calculation. The potential energy curves (Figs. 3.15 and 3.16) were calculated along the direction of the C6–N bond dissociation (i.e. from the left to right in Figs. 3.15 and 3.16), although the cyclic bond formation proceeds in the opposite direction. Complexity in finding an appropriate conformational isomer was avoided by this opposite calculation.

The obtained activation barriers (the energy differences between the transition state and the initial state) and the cyclic bond formation energies (the energy differences between the final state and the initial state) are shown in Table 3.2. All the cyclic bond formation energies were non-negative, indicating that these cyclic bond formations must be followed by intramolecular proton rearrangements to complete the overall cyclization process. Introducing α-carboxyl group and *N*-methyl group resulted in decreased activation barriers and bond formation energies, indicating enhanced nucleophilicity. In the previous report [1], a decrease in the basicity of the amino group was mentioned as the cause of the increased cyclization rate with the α-carboxyl group. Our calculations revealed that the α-carboxyl group not only reduces the basicity of the amino group but also increases the nucleophilicity. In addition, our results showed that the six-membered ring formation (a′)–(d′) requires a lower activation energy than that for the five-membered ring formation (a)–(d). To determine the factors affecting the activation barrier, we analyzed the structure of the transition state. As shown in Table 3.3, the dihedral angle θ_1 in the five-membered ring formation showed a larger variation than that for the six-membered ring formation to form the transition state. In other words, the six-membered ring formation requires a less significant distortion in the side chain conformation, giving a lower activation barrier. It can also be noted that previous studies found an involvement of six-membered spirocyclic species (resulting from nucleophilic attack on 1-carbon) as unstable intermediates in the cyclization [1]. Thus, the rate for six-membered ring formations can be complicated because such spirocyclization would hamper the normal cyclization (i.e. nucleophilic attack on C6).

Table 3.3 Change in dihedral angles during C6–N cyclic bonding from initial state to transition state ($\Delta\theta_1^{\ddagger}$, $\Delta\theta_2^{\ddagger}$, and $\Delta\theta_3^{\ddagger}$)

Label	*o*-Quinone	$\Delta\theta_1^{\ddagger}$ (deg.)	$\Delta\theta_2^{\ddagger}$ (deg.)	$\Delta\theta_3^{\ddagger}$ (deg.)
(a)	Dopaminequinone	−31.7	−51.4	–
(b)	Dopaquinone	−23.9	−23.2	–
(c)	*N*-Methyl-dopaminequinone	−50.2	12.3	–
(d)	*N*-Formyl-dopaminequinone	−36.3	−3.6	–
(a′)	*Homo*-dopaminequinone	−23.8	−6.9	−22.72
(b′)	*Homo*-dopaquinone	−15.1	−5.0	−13.42
(c′)	*N*-Methyl-*homo*-dopaminequinone	−19.6	−4.7	20.21
(d′)	*N*-Formyl-*homo*-dopaminequinone	−27.9	−6.6	24.28

Next, we discuss the effects of α-carboxylation, *N*-methylation, and *N*-formylation of dopaminequinone on the potential energy profile along the cyclic bond formation. The C6–N bond formation proceeds as the amino lone pair orbital (mainly appeared as the HOMO) spreads toward the benzene ring (Fig. 3.17). In this alteration of the orbital morphology, the amino lone pair is considered to move to the 6-carbon, making the π electron density polarized toward a quinone oxygen, 4-oxygen. Such charge transfers were observed for all the cases (Fig. 3.18). Therefore, the nucleophilicity can be enhanced when the energy level of electron-donating orbitals at the amino group such as the HOMO is up-shifted toward the vacuum level. In order to confirm this behavior, the activation barriers and the cyclic bond formation energies of (a)–(d) were plotted with respect to the HOMO levels (Fig. 3.19). The resulting plots demonstrate negative correlations between the activation/reaction energy and the HOMO level, confirming our hypothesis. It should be noted that the *N*-formylated cases were out of the linear relationship observed for the other cases. This can be explained by considering the presence of N–C π bond between the amino and the formyl group that must be broken with an additional energy during the cyclic bond formation. As shown in the inset of Fig. 3.19, we also found a linear relation between the activation energies and the cyclic bond formation energies, indicating that both are determined by the nucleophilicity of the amino group. The relatively high HOMO level of (b) dopaquinone and (c) *N*-methyl-dopaminequinone compared with the others would be due to an antibonding interaction between the amino lone pair orbital and the carboxyl π orbital, as can be seen in Figs. 3.17 and 3.20.

From the above discussion, we have pointed out that different substituent structures result in different HOMO levels, affecting the nucleophilicity. An up-shifted HOMO level corresponds to an increased electron-donating ability from the lone pair orbital, and then an enhanced nucleophilicity. To directly investigate the nucleophilicity, we calculated the condensed-to-atom Fukui indices for the *o*-quinone side chains of (a)–(d). Table 3.4 lists the calculated results. Since Fukui function is defined as the functional derivative of the (electronic) chemical potential with respect to the external potential, a highly exo-energetic electron release can occur at the site having

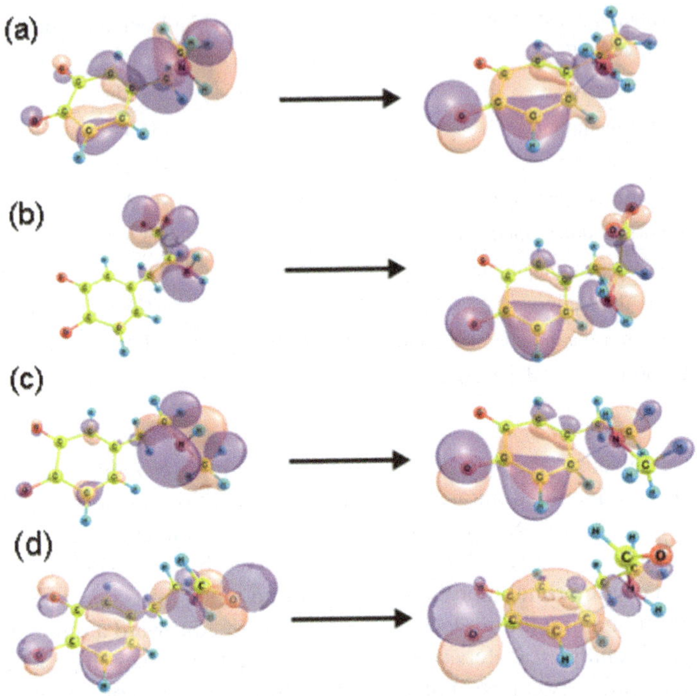

Fig. 3.17 Isosurface plots for HOMOs of **a** dopaminequinone, **b** dopaquinone, **c** *N*-methyl-dopaminequinone, and **d** *N*-formyl-dopaminequinone. (Left) Initial state and (Right) transition state of C6–N bond formation. Reprinted (with minor modification) from Ref. [22] with permission from Wiley

a high Fukui index (right derivative, f^-). As shown in Table 3.4, when summed up over all atoms, dopaquinone and *N*-methyl-dopaminequinone showed high Fukui indices, whereas the *N*-formylated side chain has a lower value.

Finally, we investigated the C6–O cyclic bond formation of RD-quinone. RD-quinone cyclizes to form a six-membered ring. As an example of the cyclized structure, we initially considered an oxonium intermediate, in which the hydroxyl O atom presents in three valencies due to the additional cyclic bonding with 6-carbon. However, we found that this oxonium cyclic structure is not stable. When this structure was relaxed by geometrical optimization, spontaneous C6–O dissociation occurred to form the uncyclized RD-quinone (Fig. 3.21). In other words, RD-quinone cannot undergo cyclization without protonation and/or deprotonation at the electroneutral condition. As possible cyclic structures, we found a hydroxyl deprotonated structure and an O4-protonated structure (Fig. 3.22). Note that, in the case of dopaquinone, the $-NH_3^+$ deprotonation is necessary for cyclization, as mentioned above. Therefore, we considered that the hydroxyl deprotonation is involved as the initial step of cyclization. Approximately speaking, a protonated amino group ($-NH_3^+$) in α-amino acids has a pK_a of 9, whereas an alcoholic hydroxyl group

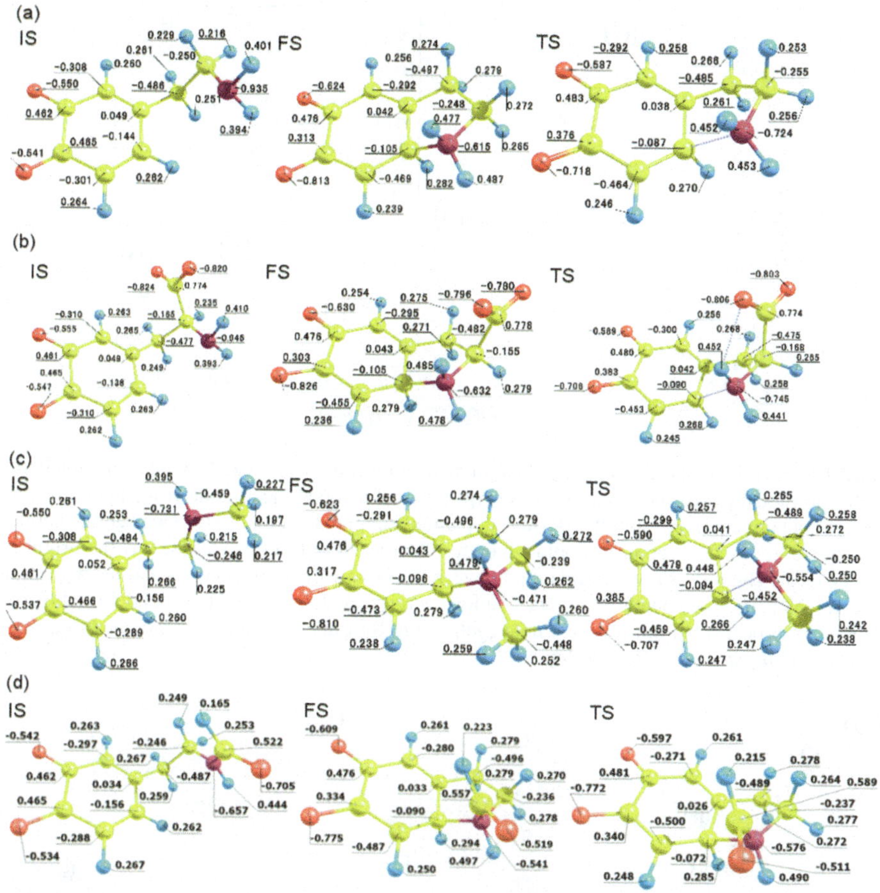

Fig. 3.18 Atomic charge (natural charge) distributions for **a** dopaminequinone, **b** dopaquinone, **c** *N*-methyl-dopaminequinone, and **d** *N*-formyl-dopaminequinone. (IS) Initial state, (FS) final state, and (TS) transition state of C6–N bond formation

(−OH) has a higher pK_a of 15. Due to this difference in pK_a, RD-quinone would be less reactive toward cyclization compared to dopaquinone. This explains a slower cyclization of RD-quinone. Nevertheless, it should be also noted that the necessity of deprotonation and/or protonation does not mean that cyclization is impossible. In our recent investigation using an extended model, we reexamined RD-quinone cyclization in comparison with dopaminequinone [34]. By considering a simultaneous proton rearrangement from the hydroxyl (or the amino) group to 4-oxygen, we found that RD-quinone can cyclize with a moderate activation energy, which is still higher than that of dopaminequinone.

At the electroneutral condition, the amino and the hydroxyl group showed a significant difference in the reactivity toward cyclization. This can be originated from the difference in energy level of the lone pair orbital, which mainly appears

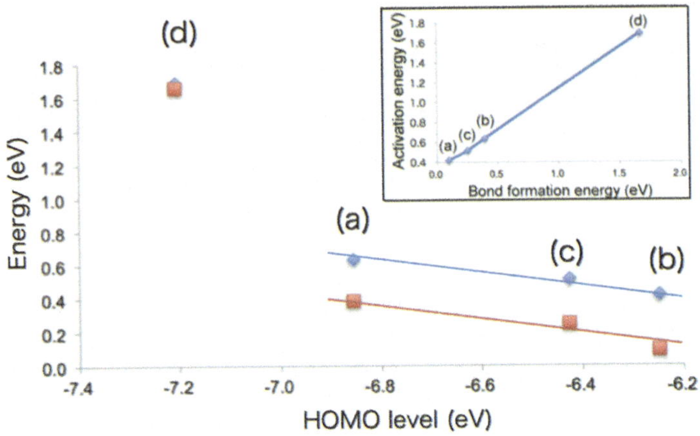

Fig. 3.19 Correlation between bond formation/activation energies for C6–N bond formation in the cyclization and the corresponding HOMO energy levels of **a** dopaminequinone, **b** dopaquinone, **c** *N*-methyl-dopaminequinone, and **d** *N*-formyl-dopaminequinone. Diamonds show activation energies and square boxes show bond formation energies. The vacuum level was chosen as the origin of energy level. The inset shows correlation between the bond formation energies and the activation energies. Note that the bond formation/activation energies are defined for C6–N bond formation (an elementary process) but not for the whole cyclization. Reprinted (with minor modification) from Ref. [22] with permission from Wiley

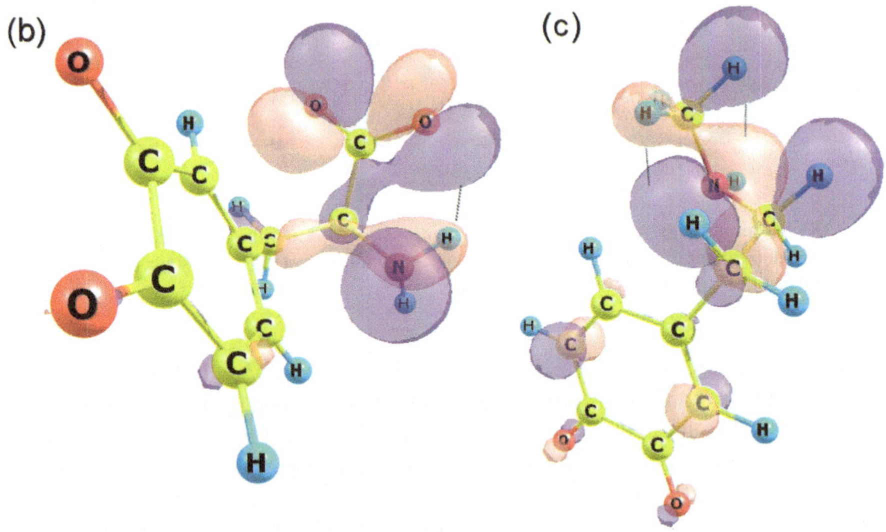

Fig. 3.20 Isosurface plots for HOMOs of **b** dopaquinone and **c** *N*-methyl-dopaminequinone with dotted lines for emphasizing anti-bonding orbital interaction (out-of-phase overlapping between two wavefunctions) inside the hydrocarbon side chain (including amino group)

Table 3.4 Condensed-to-atom Fukui indices for side chains of dopaminequinone analogs (left derivative f^+ and right derivative f^-)

Label	*o*-Quinone	Atom/Group	f^+	f^-
(a)	Dopaminequinone	Amino N	−0.01	0.33
(a)	Dopaminequinone	Whole side chain[a]	0.05	0.56
(b)	Dopaquinone	Amino N	−0.01	0.26
(b)	Dopaquinone	Carboxyl O	0.01	0.19
(b)	Dopaquinone	Carboxyl O	0.01	0.33
(b)	Dopaquinone	Whole side chain[a]	0.04	0.91
(c)	*N*-Methyl-dopaminequinone	Amino N	0.00	0.45
(c)	*N*-Methyl-dopaminequinone	*N*-Methyl C	0.00	−0.05
(c)	*N*-Methyl-dopaminequinone	Whole side chain[a]	0.04	0.80
(d)	*N*-Formyl-dopaminequinone	Amino N	−0.01	0.02
(d)	*N*-Formyl-dopaminequinone	*N*-Formyl C	0.00	0.02
(d)	*N*-Formyl-dopaminequinone	*N*-Formyl O	0.00	0.00
(d)	*N*-Formyl-dopaminequinone	Whole side chain[a]	0.05	0.11

[a]Condensed-to-atom Fukui indices were integrated for all the side chain atoms

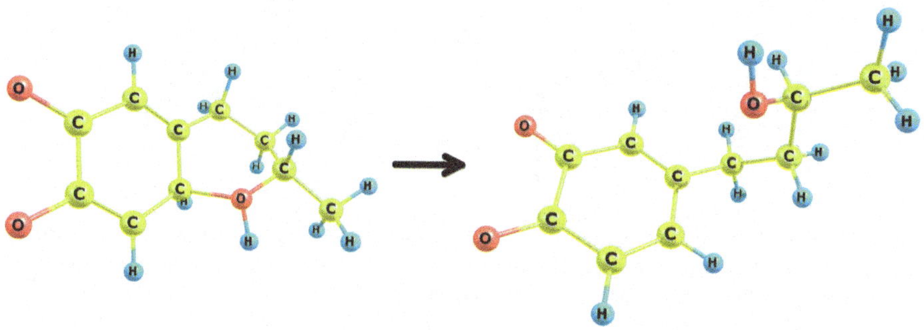

Fig. 3.21 (Left) Unstable cyclic intermediate resulting from C6–O bond formation of RD-quinone. Structural relaxation results in spontaneous change into (Right) uncyclized structure. Reprinted from Ref. [21] with permission from Physical Society of Japan

as HOMO. Dopaquinone and RD-quinone has the HOMO level of −6.0 and −7.3 eV, respectively. Therefore, the hydroxyl electrons in RD-quinone are harder to be removed compared to the amino electrons in dopaquinone. This makes the RD-quinone cyclization, which occurs with a charge transfer from the hydroxyl group, energetically less preferred. However, once the hydroxyl group is deprotonated, the lone pair level is up-shifted toward the vacuum level due to the absence of the proton coordination, and then the cyclic bond formation can take place. When 4-oxygen is protonated, the π electron energy levels in the benzene ring is down-shifted, and then they become aggressive in accepting electrons from the hydroxyl group. These

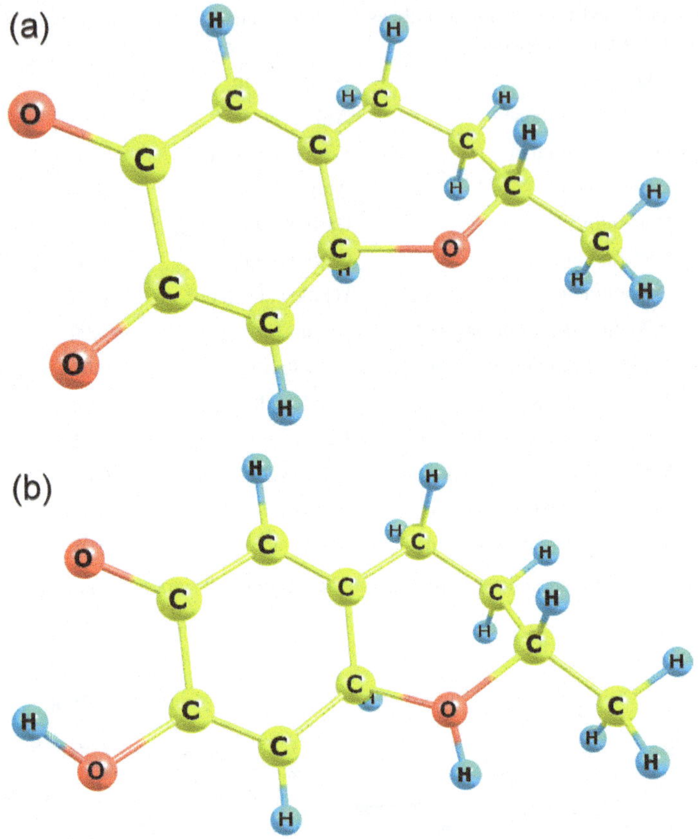

Fig. 3.22 Stable cyclic intermediates resulting from C6–O bond formation of **a** hydroxy-deprotonated and **b** O4-protonated RD-quinone. Reprinted (with minor modification) from Ref. [21] with permission from Physical Society of Japan

changes in the electronic states explain the stability of the hydroxyl deprotonated and O4-protonated cyclic structures (Fig. 3.22).

As possible factors promoting RD-quinone cyclization, we considered *O*-methylation of the hydroxyl group and carboxylation at the position adjacent to the hydroxyl group, based on the results for dopaminequinone analogs. By introducing these substituents, the HOMO levels were up-shifted (RD-quinone: −7.4 eV, *O*-methylated derivative: −7.1 eV, and carboxylated derivative: −6.4 eV). As shown in Fig. 3.23, the *O*-methylated RD-quinone cannot form a cyclic structure, whereas the carboxylated RD-quinone exhibits a stabilization during the reaction. At the point showing a non-smooth change in the potential energy curve for the carboxylated case, a proton transfer between the hydroxyl group and the carboxyl group was observed. This proton transfer may contribute to stabilization of the oxonium cyclic structure. Unlike the cases in dopaminequinone analogs, methylation was not effective enough

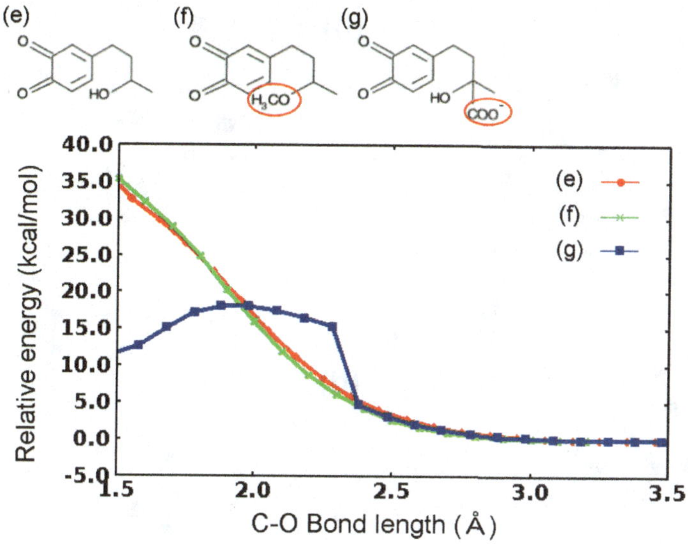

Fig. 3.23 Potential energy curves for C6–O bond formation of **e** RD-quinone, **f** 4-(2,3-quinonyl)-2-methoxybutane, and **g** 4-(2,3-quinonyl)-2-carboxybutanol

to stabilize the unstable oxonium structure. Here, we predicted the effects of carboxylation of RD-quinone on its cyclization. By promoting RD-quinone cyclization with chemical modifications, it is expected to indirectly inhibit the thiol binding, and then reduce the cytotoxicity.

3.5 Binding of Cysteine with Dopaquinone

Here, we introduce a mechanistic study on cysteine binding with dopaquinone. Initially, we tried to find a cysteine-bound structure using the non-deprotonated sulfhydryl group (−SH). However, no stable bound-structures were found, as manifested by spontaneous dissociation upon geometrical optimization. Accordingly, we considered that the sulfhydryl deprotonation is the initial step for the cysteine binding, and we solely used the deprotonated cysteine thiolate ion (Cys–S$^-$) for the calculations. This is consistent with the base-catalyzed kinetics reported previously [17, 18].

We obtained five Cys–S$^-$-bound structures (a–e) as shown in Fig. 3.24. The binding energies calculated are shown in Table 3.5. As mentioned above, the C5- and C2-adducts but not C6-adducts have been experimentally found as the major products [2, 14–16]. Nevertheless, our results show that the C2-bound structure (b) is less energetically favorable than the C6-bound case (c). As shown in Table 3.5, cysteine preferred the C3–C4 bridge (d) more than the C5 (a) and C2 (b). This C3–C4-bound structure (d) has a relatively long C–S bond length (2.75 Å of C3–S

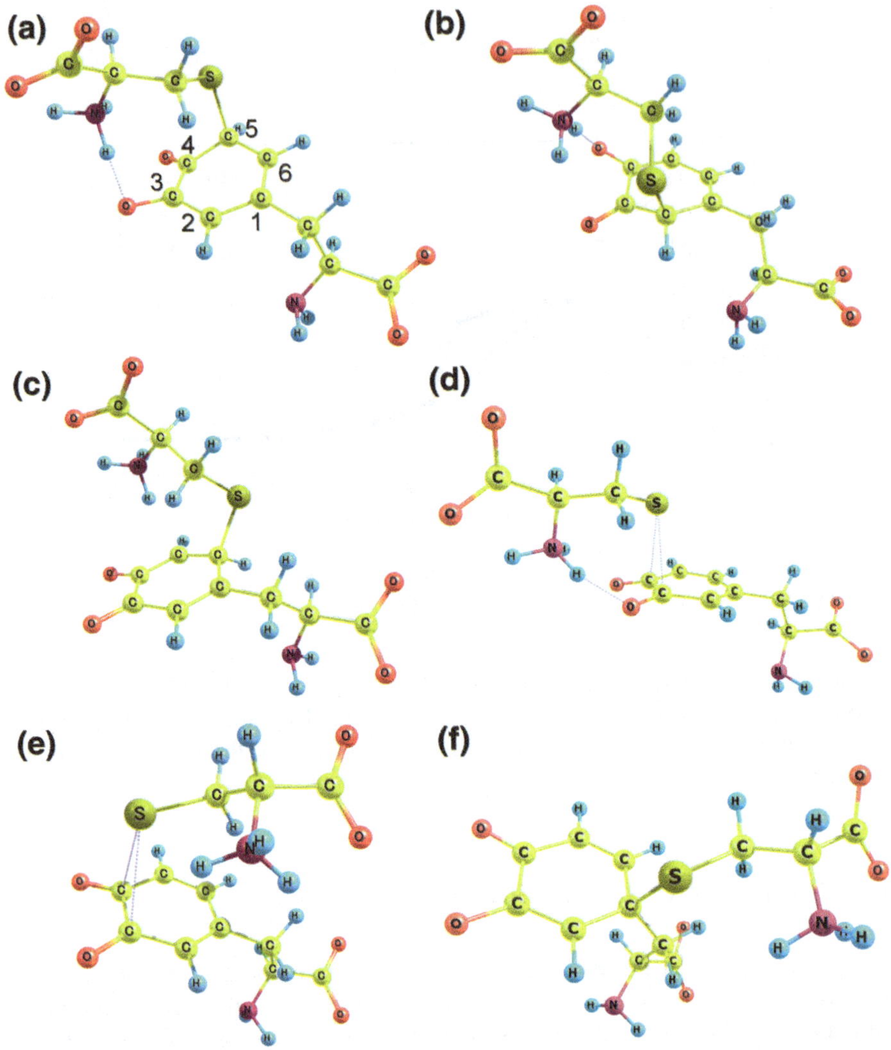

Fig. 3.24 Bond formation between S in cysteine thiolate ion Cys-S$^-$ and **a** C5, **b** C2, **c** C6, **d** C3–C4 bridge, and **e** C1 in dopaquinone

Table 3.5 Binding energies of cysteine thiolate Cys−S$^-$ to dopaquinone

Label	Binding site	Binding energy (kcal/mol)
A	C5	5.8
B	C2	4.5
C	C6	5.2
D	C3−C4	9.8
E	C1	4.0

bond length), indicating a less ordinary covalent bonding unlike the other cases; for instance, the C5-bound structure (a) shows 1.91 Å of C5–S bond length. The unusual covalent nature of the C3–C4-bound structure (d) also manifests in a relatively small geometrical alteration upon binding unlike the other cases (a, b, c, and e) exhibiting a *sp*^{2}-to-*sp*^{3} (planar-to-pyramidal) structural change during the reaction.

As a further possible process, we investigated a reaction path of Cys–S$^-$ migration from C3–C4 bridge to C5. Then, we considered two coordinates *Z* and *D*, as defined in Fig. 3.25, to describe this migration. *Z* is the height of Cys–S$^-$ as measured from C3, and *D* increases as Cys–S$^-$ migrates along the perimeter of the benzene ring. The calculated potential energy surface along the two degrees of freedom *Z* and *D* is shown in Fig. 3.26. Note that the benzene ring carbon atoms, and all the Cys–S$^-$ atoms were fixed, and the other degrees of freedom were relaxed during the calculation. Our result shows the absence of activation barrier for the binding onto C3–C4 bridge, while the potential energy increases around C5. From this finding, it would be advisable to consider C3–C4 bridge but not C5 as the initial binding site.

Based on the above findings, we hypothesized that Cys–S$^-$ is initially bound onto the C3–C4 bridge, and then migrates to C5 or C2, followed by several conversions

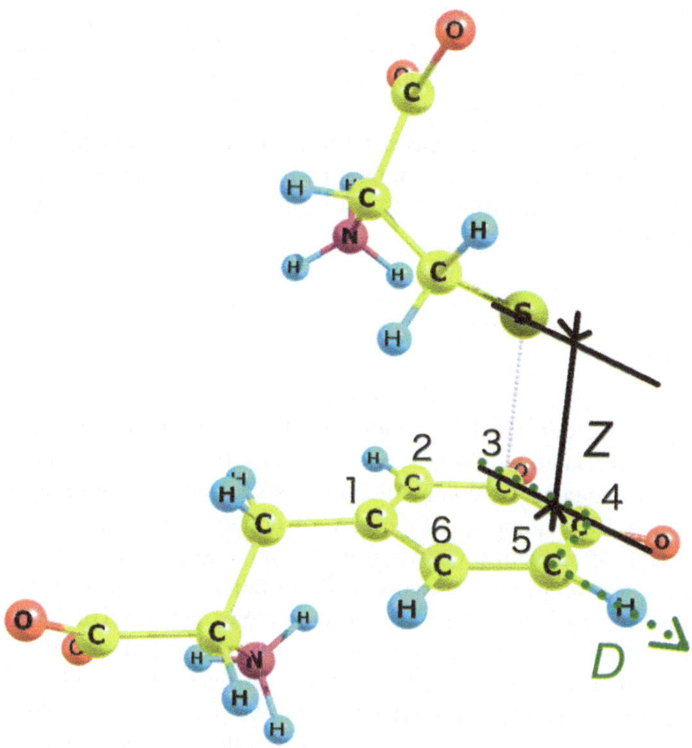

Fig. 3.25 Two coordinates *Z* and *D* for bond formation between S in cysteine thiolate ion Cys-S$^-$ and C5 in dopaquinone

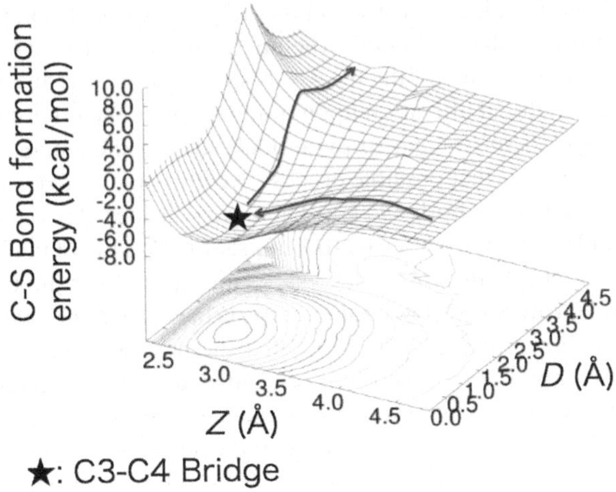

Fig. 3.26 Potential energy surface along *Z* and *D* (defined in Fig. 3.25). The arrow shows an approaching route for Cys-S⁻ on C5 in dopaquinone

to be a more stable structure. After the migration to C5 or C2, this reaction must be completed by subsequent proton rearrangements to form the product cysteinyl-dopa. From the C5– (or C2–) bound structure, the amino group in Cys–S⁻ can release its proton into the hydrogen-bonded O3 (or O4), and then further proton rearrangement from C5 (or C2) to O4 (or O3) gives rise to 5-*S*-cysteinyldopa (or 2-*S*-cysteinyldopa). As the representative case, we calculated the energy diagram for the 5-*S*-cysteinyldopa formation based on the hypothesized scheme. The obtained results are shown in Fig. 3.27. As can be seen in the drastic decrease in energy, the final proton rearrangement from C5 to O4 is an important process that makes this reaction system irreversible.

3.6 Summary

In this chapter, we described our studies on the competitive reactions involving dopaquinone and related *o*-quinones, namely cyclization and thiol binding.

First, we investigated the competitive effects of thiol binding on cyclization of dopaquinone and RD-quinone. In the case of dopaquinone, the thiol-bound state became unstable after cyclization, whereas RD-quinone could bind thiols even after cyclization. Since thiol binding involves redistribution of electronic charge to the *o*-quinone, the binding energy becomes lower as the LUMO level of the *o*-quinone is up-shifted toward the vacuum level. In our calculation, both dopaquinone and RD-quinone showed an up-shifted LUMO level by cyclization. Especially in the case of RD-quinone, this up-shift was remarkably high. We pointed out that this difference in the LUMO level shift is due to the difference in the covalent and ionic character of the C6–N bond and the C6–O bond.

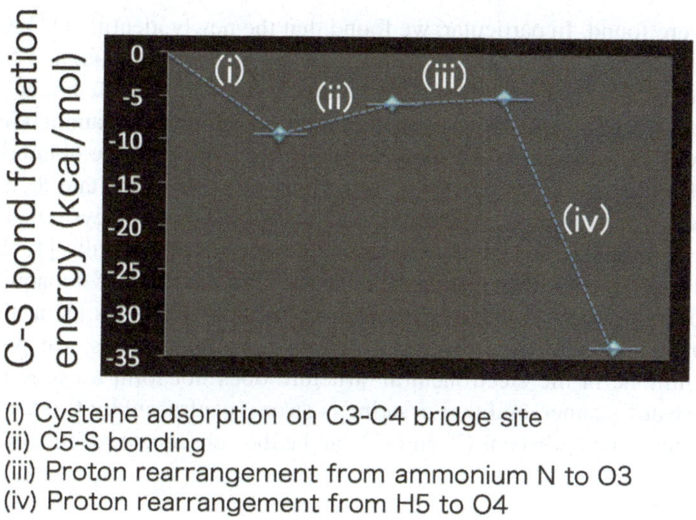

Fig. 3.27 Energy diagram for binding of Cys–S$^-$ on dopaquinone. (i) Recruitment of Cys–S$^-$ on C3–C4 bridge, (ii) C5–S bond formation, (iii) Proton transfer from ammonium group in Cys–S$^-$ to O3, and (iv) Proton transfer from ammonium group in C5 to O4. Note that transition states are not included

Next, we investigated the initial process of *o*-quinone cyclization, namely C6–N or C6–O cyclic bond formation. For the C6–N cyclic bond formation, dopaminequinone, dopaquinone, *N*-methyl-dopaminequinone, *N*-formyl-dopaminequinone (five-membered ring formation), and those with methylene-inserted hydrocarbon chains (six-membered ring formation) were selected for calculations. For the C6–N cyclic bond formation, we focused on cyclization of RD-quinone. As a result, it was revealed that the introduction of α-carboxyl group and *N*-methyl group has the effect of lowering the activation barrier for C6–N cyclic bond formation. On the other hand, the introduction of *N*-formyl group had the opposite effect of raising the activation barrier. Comparing with the methylene-inserted cases, the six-membered ring formation showed a lower activation barrier than that of the five-membered ring formation. The lone pair orbital of the amino group is partially included in the HOMO, and then the lone pair charge is donated to the benzene ring. Therefore, the closer this level is to the vacuum level, the greater the nucleophilicity will become. Here, we showed that the activation barrier and the reaction energy (bond formation energy) have a linear relationship, and also correlate with the HOMO level. As a cause of higher HOMO level, we emphasized an anti-bonding orbital interaction induced by α-carboxylation and *N*-methylation. Unlike the cases of C6–N bond formation, RD-quinone did not form C6–O cyclic bond at the electroneutral condition. Instead, we showed that the hydroxyl deprotonation can be the initial process for cyclization of RD-quinone.

Finally, we investigated the binding of cysteine to dopaquinone. As a result, five binding sites for cysteine thiolate ion Cys–S$^-$, namely C5, C2, C6, C3–C4 bridge, and C1, were found. In particular, we found that the newly identified C3–C4 bridge is the energetically favorable site more than the previously thought C5 and C2. The C6-bound structure showed higher binding energy than the C2-bound structure, indicating the experimentally observed preference of C2 over C6 cannot be explained by the energetic stability of these bound structures. Based on the obtained results, we proposed that cysteine is initially bound onto C3–C4 bridge, and then migrates to the adjacent C5 or C2 as the mechanism for the initial step of cysteine binding.

These findings are in agreement with the experimental results [1, 2, 7], and explains the tendency of cyclization/thiol-binding competition of dopaquinone and RD-quinone [2, 7] and cyclization rates of dopaquinone analogs [1] at the atomic level. Furthermore, beyond the extent of experimental findings, we pointed out that RD-quinone in the electroneutral structure does not form a cyclic bond in a straightforward manner, and that cysteine is strongly attracted by C3–C4 bridge in dopaquinone so that adjacent C5 and C2 can be the subsequent reaction sites.

References

1. E.J. Land, C.A. Ramsden, P.A. Riley, *ortho*-Quinone amines and derivatives: the influence of structure on the rates and modes of intramolecular reaction. Arkivoc. **xi**, 23–36 (2007)
2. S. Ito, A Chemist's view of melanogenesis. Pigment Cell Res. **16**, 230–236 (2003)
3. W.D. Bush, J. Garguilo, F.A. Zucca, A. Albertini, L. Zecca, G.S. Edwards, R.J. Nemanich, J.D. Simon, The surface oxidation potential of human neuromelanin reveals a spherical architecture with a pheomelanin core and a eumelanin surface. Natl. Acad. Sci. **103**, 14785–14789 (2006)
4. S. Ito, Encapsulation of a reactive core in neuromelanin. Proc. Natl. Acad. Sci. **103**, 14647–14648 (2006)
5. S. Ito, T. Kato, K. Fujita, Covalent binding of catechols to proteins through the sulphydryl group. Biochem. Pharmacol. **37**, 1707–1710 (1988)
6. K. Hasegawa, S. Ito, S. Inoue, K. Wakamatsu, H. Ozeki, I. Ishiguro, Dihydro-1,4-benzothiazine-6,7-dione, the ultimate toxic metabolite of 4-*S*-cysteaminylphenol and 4-*S*-cysteaminylcatechol. Biochem. Pharmacol. **53**, 1435–1444 (1997)
7. S. Ito, M. Ojika, T. Yamashita, K. Wakamatsu, Tyrosinase-catalyzed oxidation of rhododendrol produces 2-methylchromane-6,7-dione, the putative ultimate toxic metabolite: implications for melanocyte toxicity. Pigment Cell Melanoma Res. **27**, 744–753 (2014)
8. A. Thompson, E.J. Land, M.R. Chedekel, K.V. Subbarao, T.G. Truscott, A pulse radiolysis investigation of the oxidation of the melanin precursors 3,4-dihydroxyphenylalanine (dopa) and the cysteinyldopas. Biochim. Biophys. Acta **843**, 49–57 (1985)
9. M.D. Hawley, S.V. Tatawawadi, S. Piekarski, R.N. Adams, Electrochemical studies of the oxidation pathways of catecholamines. J. Am. Chem. Soc. **89**, 447–450 (1967)
10. J. Borovansky, R. Edge, E.J. Land, S. Navaratnam, S. Pavel, C.A. Ramsden, P.A. Riley, N.P.M. Smit, Mechanistic studies of melanogenesis: the influence of *N*-substitution on dopamine quinone cyclization. Pigment Cell Melanoma Res. **19**, 170–178 (2006)
11. T.E. Young, J.R. Griswold, M.H. Hulbert, Melanin. I. Kinetics of the oxidative cyclization of dopa to dopaquinone. J. Org. Chem. **39**, 1980–1982 (1974)
12. J. Cabanes, F. García-Cánovas, J.A. Lozano, F. García-Carmona, A kinetic study of the melanization pathway between L-tyrosine and dopachrome. Biochim. Biophys. Acta **923**, 187–195 (1987)

13. F. García-Carmona, F. García-Cánovas, J.L. Iborra, J.A. Lozano, Kinetic study of the pathway of melanization between L-dopa and dopachrome. Biochim. Biophys. Acta **717**, 124–131 (1982)
14. S. Ito, K. Wakamatsu, Chemistry of mixed melanogenesis-pivotal roles of dopaquinone. Photochem. Photobiol. **84**, 582–592 (2008)
15. S. Ito, G. Prota, A facile one-step synthesis of cysteinyldopas using mushroom tyrosinase. Experimentia **33**, 1118–1119 (1977)
16. G. Prota, Recent advances in the chemistry of melanogenesis in mammals. J. Invest. Dermatol. **75**, 122–127 (1980)
17. X. Huang, R. Xu, M.D. Hawley, T.L. Hopkins, K.J. Kramer, Electrochemical oxidation of *N*-acyldopamines and regioselective reactions of their quinones with *N*-acetylcysteine and thiourea. Arch. Biochem. Biophys. **352**, 19–30 (1998)
18. R. Xu, X. Huang, K.J. Kramer, M.D. Hawley, Characterization of products from the reactions of dopamine quinone with *N*-acetylcysteine. Bioorg. Chem. **24**, 110–126 (1996)
19. G.N.L. Jameson, J. Zhang, R.F. Jameson, W. Linert, Kinetic evidence that cysteine reacts with dopaminoquinone via reversible adduct formation to yield 5-cysteinyl-dopamine: an important precursor of neuromelanin. Org. Biomol. Chem. **2**, 777–782 (2004)
20. R. Kishida, H. Kasai, S.M. Aspera, R.L. Arevalo, H. Nakanishi, Branching reaction in melanogenesis: the effect of intramolecular cyclization on thiol binding. J. Electron. Mater. **46**, 3784–3788 (2017)
21. R. Kishida, H. Kasai, S.M. Aspera, R.L. Arevalo, H. Nakanishi, Density functional theory-based first principles calculations of rhododendrol-quinone reactions: preference to thiol binding over cyclization. J. Phys. Soc. Jpn. **86**, 024804-1–5 (2017)
22. R. Kishida, A.G. Saputro, R.L. Arevalo, H. Kasai, Effects of introduction of α-carboxylate *N*-methyl and *N*-formyl groups on intramolecular cyclization of *o*-quinone amines: density functional theory-based study. Int. J. Quant. Chem. **117**, e25445-1-9 (2017)
23. P. Hohenberg, W. Kohn, Inhomogeneous electron gas. Phys. Rev. **136**, B864–B871 (1964)
24. W. Kohn, L.J. Sham, Self-consistent equations including exchange and correlation effects. Phys. Rev. **140**, A1133–A1138 (1965)
25. A.D. Becke, Density-functional thermochemistry. III. The role of exact exchange. J. Chem. Phys. **98**, 5648–5652 (1993)
26. C. Lee, W. Yang, R.G. Parr, Development of the Colle-Salvetti correlation-energy formula into a functional of the electron density. Phys. Rev. B **37**, 785–789 (1988)
27. M.J. Frisch, G.W. Trucks, H.B. Schlegel, G.E. Scuseria, M.A. Robb, J.R. Cheeseman, G. Scalmani, V. Barone, B. Mennucci, G.A. Petersson, H. Nakatsuji, M. Caricato, X. Li, H.P. Hratchian, A.F. Izmaylov, J. Bloino, G. Zheng, J.L. Sonnenberg, M. Hada, M. Ehara, K. Toyota, R. Fukuda, J. Hasegawa, M. Ishida, T. Nakajima, Y. Honda, O. Kitao, H. Nakai, T. Vreven, J.A. Montgomery, Jr., J.E. Peralta, F. Ogliaro, M. Bearpark, J.J. Heyd, E. Brothers, K.N. Kudin, V.N. Staroverov, R. Kobayashi, J. Normand, K. Raghavachari, A. Rendell, J.C. Burant, S.S. Iyengar, J. Tomasi, M. Cossi, N. Rega, J.M. Millam, M. Klene, J.E. Knox, J.B. Cross, V. Bakken, C. Adamo, J. Jaramillo, R. Gomperts, R.E. Stratmann, O. Yazyev, A.J. Austin, R. Cammi, C. Pomelli, J.W. Ochterski, R.L. Martin, K. Morokuma, V.G. Zakrzewski, G.A. Voth, P. Salvador, J.J. Dannenberg, S. Dapprich, A.D. Daniels, Ö. Farkas, J.B. Foresman, J.V. Ortiz, J. Cioslowski, D.J. Fox, Gaussian 09, Revision C. 01, Gaussian, Inc., Wallingford CT (2009)
28. J.P. Foster, F. Weinhold, Natural hybrid orbitals. J. Am. Chem. Soc. **102**, 7211–7218 (1980)
29. J. Tomasi, B. Mennucci, R. Cammi, Quantum mechanical continuum solvation models. Chem. Rev. **105**, 2999–3093 (2005)
30. J.L. Pascual-Ahuir, E. Silla, I. Tuñon, GEPOL: an improved description of molecular surfaces. III. A new algorithm for the computation of a solvent-excluding surface. J. Comput. Chem. **15**, 1127–1138 (1994)
31. R.G. Parr, W. Yang, Density functional approatch to the frontier-electron theory of chemical reactivity. J. Am. Chem. Soc. **106**, 4049–4050 (1984)
32. P. Fuentealba, P. Pérez, R. Contreras, On the condensed Fukui function. J. Chem. Phys. **113**, 2544–2551 (2000)

33. C. Peng, H.B. Schlegel, Combining synchronous transit and quasi-newton methods for finding transition states. Isr. J. Chem. **33**, 449–454 (1993)
34. R. Kishida, H. Kasai, Cyclic bond formation of rhododendrol-quinone and dopamine-quinone: effects of proton rearrangement. J. Phys. Soc. Jpn. **87**, 084802-1–5 (2018)

Chapter 4
Concluding Remarks and Future Perspectives

Abstract Summarizing our findings, we give concluding remarks and future perspectives. One of next challenges will be to unveil quantum effects of atomic nuclei in melanogenesis. While the classical description of atomic nuclei is sufficient for most scenarios, certain phenomena require quantum mechanical description of both atomic nuclei and electrons.

Keywords Computational materials design (CMD®) · Quantum effects

In this book, we introduced melanin chemistry studies with the emphasis of computational approach. With the aid of quantum chemical calculation, we aimed to understand branched reactions involved in melanogenesis, namely dopachrome conversion and cyclization/thiol binding of dopaquinone analogues. The former reaction, dopachrome conversion produces eumelanin monomers DHI and DHICA, whereas the latter reaction cyclization/thiol binding of dopaquinone analogs results in the switching between eumelanogenesis and pheomelanogenesis.

As described in Chap. 2, the computational studies of dopachrome conversion found important factors affecting the selective formation of DHI and DHICA. The branch into DHI and DHICA formation, respectively, results from decarboxylation and deprotonation from α-carbon of dopachrome. These reaction modes are controlled by the protonated/deprotonated state of the quinonoid group that is located distant from α-carbon. This is the result that the characteristics of π-conjugated system are clearly exhibited. Integrating the obtained results, the catalytic effects of basic pH and Cu(II) experimentally observed can be explained based on two aspects: promotion of the rate-determining step and protection of quinonoid group from protonation. Thus, the computational approach provides the atomic-scale understanding of dopachrome conversion. It is expected that these findings are used to predict the relation between the synthetic condition and the eumelanin composition, namely the DHI/DHICA ratio.

As described in Chap. 3, the computational studies of cyclization and thiol binding for dopaquinone and structurally similar *o*-quinones revealed key factors affecting the reactivity toward the two reaction modes. Cyclization and thiol binding

R. Kishida et al., *Melanin Chemistry Explored by Quantum Mechanics*,
https://doi.org/10.1007/978-981-16-1315-9_4

are in a competition, affecting the melanin composition by determining the eumelanin/pheomelanin ratio. Since both the reactions proceed with charge redistribution, the HOMO and LUMO level of *o*-quinone are important factors affecting the reactivity.

To demonstrate the competition, we showed that binding of thiols become unfavorable upon cyclization. Especially, in the case of dopaquinone, the cyclized product dopachrome showed a significantly reduced binding energy to thiols with a highly up-shifted LUMO level by cyclization. In contrast, RD-quinone can bind thiols even after cyclization with a less up-shifted LUMO level by cyclization.

From an investigation on cyclization of *o*-quinone and substituent effects, it was found that α-carboxylation and *N*-methylation enhances the nucleophilicity of the side chain. It was pointed out that the amino lone pair electrons became more unstable by introducing these substituents, and results in a more aggressive charge donation by means of nucleophilic attack. In the case of RD-quinone, cyclic C−O bond formation requires hydroxyl deprotonation (or quinonic protonation), indicating a less preference for cyclization over thiol binding. Considering several cytotoxic effects of thiol binding reported so far, this preference for thiol binding of RD-quinone may be a factor contributing to the cytotoxicity.

Aiming to understand the binding mechanism of cysteine to dopaquinone, the energetic preference of cysteine thiolate ion for various possible binding sites was investigated. As an important finding, C3−C4 bridge site but not C5 and C2 was identified as the most stable site. Although the C6-adduct has never been reported as the major product, our calculated results showed that the binding at C2 is not energetically stable than at C6. Therefore, an alternative explanation other than straightforward energetics augment is necessary. As the mechanism for the initial step for cysteine binding, it was proposed that cysteine thiolate is initially bound onto C3−C4 bridge, and then migrates to the adjacent sites C5 and C2. Recently, we have proposed a mechanism of the cysteine addition reaction to form cysteinyldopa via a thiolate-attacked intermediate at C3−C4 bridge [1].

While the classical description of atomic nuclei is sufficient for most scenarios, certain phenomena require quantum mechanical description to fully understand the phenomena. Hydrogen exhibits quantum mechanical phenomena compared to other elements, as quantum mechanical features are significantly associated with low-mass particles. From our previous studies, we have shown that quantum mechanical treatment is essential for describing the behavior of hydrogen on solid surfaces [2]. As shown in this book, proton transfer such as protonation and deprotonation proceed in melanin formation, and a quantum mechanical description of the nucleus is essential. Our preliminary results show that the quantum tunneling effect is considerable. Further research will be conducted to gain a complete understanding of melanin chemistry from the atomic nuclear and electronic world.

Melanin chemistry has been developed through multi-disciplinary collaboration. In this book, we introduced case studies on chemical reactions in melanogenesis based on CMD® approach, where the electronic states of molecules are computed from the first principles of quantum theory. From the computational studies, various factors that control the reaction were clarified and the atomic-scale understanding of the reaction mechanisms was achieved. We believe that the obtained knowledge introduced here provides a foundation for the development of melanin chemistry and related studies.

References

1. R. Kishida, S. Ito, M. Sugumaran, R.L. Arevalo, H. Nakanishi, H. Kasai, Density functional theory-based calculation shed new light on the bizarre addition of cysteine thiol to dopaquinone, Int. J. Mol. Sci. **22**, 1373 (2021)
2. H. Kasai, A.A.B. Padama, B. Chantaramolee, R.L. Arevalo, Hydrogen and hydrogen-containing molecules on metal surfaces (Springer 2020)

The manufacturer's authorised representative in the EU is Springer Nature Customer Service Centre GmbH, Europaplatz 3, 69115 Heidelberg, Germany. If you have any concerns regarding our products, please contact ProductSafety@springernature.com

Printed and bound by CPI Group (UK) Ltd, Croydon, CR0 4YY
30/09/2026
02225307-0001